Introduction to the
Design and Analysis of
Composite
Structures

An Engineer's Practical Guide
Using OptiStruct®

SECOND EDITION

Jeffrey A. Wollschlager

Published by
Jeffrey A. Wollschlager
JLab Composites
PO BOX 14609
Mill Creek, WA 98082

Introduction to the Design and Analysis of Composite Structures
An Engineer's Practical Guide
Using OptiStruct®
Second Edition

Printed by CreateSpace, An Amazon.com Company

To my wife:
Tracy
Thank you for your never ending love and support.

And my children:
Callie, Cooper
Yes, I can play now!

Contents

CONTENTS V

LIST OF FIGURES.. XI

INTRODUCTION XIII

CHAPTER 1 INTRODUCTION TO COMPOSITE MATERIALS 1

  1.1 COMPOSITE MATERIALS 1
    Particulate Composites *1*
    Laminated Composites *2*
    Fiber-Matrix Laminated Composites *2*
    Core-Stiffened Laminated Composites *5*
  1.2 MATERIAL TERMINOLOGY 5
    Microscopic *5*
    Macroscopic..*5*
    Micromechanics *5*
    Macromechanics *6*
    Homogeneous...*6*
    Heterogeneous...*6*
    Isotropic *6*
    Orthotropic..*6*
    Anisotropic *6*
  1.3 PLY CONVENTIONS 7
  1.4 LAMINATE CONVENTIONS AND DEFINITIONS 7
    Defining Laminates *8*
    Symmetric Laminates *9*
    Anti-Symmetric Laminates *9*
    Balanced Laminates *9*
    Cross-Ply Laminates *10*
    Angle-Ply Laminates *10*
    General Laminates *10*

CHAPTER 2 ELASTICITY REVIEW 11

  2.1 COORDINATE SYSTEMS 11
  2.2 STRESS...12
  2.3 STRESS INVARIANTS..14
  2.4 EQUILIBRIUM EQUATIONS...18
  2.5 STRAIN 20
  2.6 STRAIN INVARIANTS 24
  2.7 GENERALIZED HOOKE'S LAW 27
  2.8 GENERAL ELASTICITY PROBLEM 28
  2.9 VECTOR TRANSFORMATION LAW 31

2.10 2ND ORDER TENSOR TRANSFORMATION LAW 33
2.11 4TH ORDER TENSOR TRANSFORMATION LAW 34
CHAPTER 2 EXERCISES 35

CHAPTER 3 LINEAR ELASTIC MATERIAL LAWS 37

3.1 ANISOTROPIC MATERIAL LAW 37
Anisotropic Compliance Matrix.. 37
Boundary Condition #1.. 39
Boundary Condition #2.. 40
Boundary Condition #3.. 41
Boundary Condition #4.. 41
Boundary Condition #5.. 42
Boundary Condition #6.. 43
Anisotropic Stiffness Matrix 45
Anisotropic Plane Stress Compliance Matrix............................... 46
Anisotropic Plane Stress Stiffness Matrix.................................... 46
3.2 ISOTROPIC MATERIAL LAW 47
Isotropic Compliance Matrix... 47
Isotropic Stiffness Matrix.. 48
Isotropic Plane Stress Compliance Matrix................................... 49
Isotropic Plane Stress Stiffness Matrix....................................... 50
Restrictions on Isotropic Engineering Constants 51
3.3 TRANSVERSELY ISOTROPIC MATERIAL LAW 52
Transversely Isotropic Compliance Matrix................................. 52
Transversely Isotropic Stiffness Matrix...................................... 53
Transversely Isotropic Plane Stress Compliance Matrix 55
Transversely Isotropic Plane Stress Stiffness Matrix 56
Restrictions on Transversely Isotropic Engineering Constants.......... 57
3.4 ORTHOTROPIC MATERIAL LAW 58
Orthotropic Compliance Matrix.. 58
Orthotropic Stiffness Matrix... 59
Orthotropic Plane Stress Compliance Matrix 61
Orthotropic Plane Stress Stiffness Matrix 62
Restrictions on Orthotropic Engineering Constants.................... 63
3.5 GENERALIZED HOOKE'S LAW INCLUDING THERMAL STRAINS 64
CHAPTER 3 EXERCISES 66

CHAPTER 4 MICROMECHANICS OF COMPOSITE MATERIALS 69

4.1 RULE OF MIXTURES THEORY 69
Determining E_1 of a Transversely Isotropic Ply.......................... 71
Determining E_2 of a Transversely Isotropic Ply.......................... 72
Determining v_{12} of a Transversely Isotropic Ply......................... 74
Determination of v_{23} of a Transversely Isotropic Ply.................... 75

Determining G_{12} of a Transversely Isotropic Ply.............................*77*
Determining α_1 of a Transversely Isotropic Ply *79*
Determining α_2 of a Transversely Isotopic Ply..........................*80*
4.2 MODIFIED RULE OF MIXTURES THEORY..................................81
Determining E_2 using the Stress Partitioning Factor.................*82*
Determining v_{23} using the Stress Partitioning Factor................*83*
Determining G_{12} using the Stress Partitioning Factor................*84*
4.3 SUMMARY OF EQUATIONS 85
Summary of Rule of Mixtures Equations...............................*85*
Homogenized Coefficients of Thermal Expansion *85*
Summary of Modified Rule of Mixtures Equations *86*
Homogenized Coefficients of Thermal Expansion *86*
4.4 PHASE AVERAGE THEORY 87
CHAPTER 4 EXERCISES...90

CHAPTER 5 MACROMECHANICAL BEHAVIOR OF A PLY.........**93**

5.1 PLANE STRESS STIFFNESS MATRIX IN GLOBAL COORDINATES.........94
CHAPTER 5 EXERCISES...98

CHAPTER 6 CLASSICAL LAMINATION THEORY..........................**99**

6.1 LAMINATE CONVENTIONS..99
6.2 KIRCHHOFF PLATE THEORY 100
6.3 PLATE RESULTANT FORCES AND MOMENTS..........................102
6.4 ABD MATRIX OF A LAMINATED PLATE..............................104
6.6 SEQUENCE FOR SOLVING LAMINATED PLATE PROBLEMS 108
CHAPTER 6 EXERCISES...110

**CHAPTER 7 EQUIVALENT HOMOGENEOUS PLATE
CALCULATIONS**...**113**

7.1 ABD MATRIX OF A HOMOGENEOUS PLATE...........................114
7.2 EQUIVALENT IN-PLANE HOMOGENEOUS ENGINEERING CONSTANTS116
7.3 EQUIVALENT BENDING HOMOGENEOUS ENGINEERING CONSTANTS 118
7.4 EQUIVALENT HOMOGENEOUS MATERIAL MATRICES 120
7.5 EQUIVALENT IN-PLANE QUASI-ISOTROPIC LAMINATE 123
7.6 EQUIVALENT BENDING QUASI-ISOTROPIC LAMINATE 125
CHAPTER 7 EXERCISES...128

CHAPTER 8 SMEAR TECHNOLOGY **129**

8.1 STACKING SEQUENCE DEPENDENCE OF [ABD] MATRIX.................129
8.2 SMEAR TECHNOLOGY IN COMPOSITE DESIGN 133
8.3 SMEAR CORE TECHNOLOGY IN COMPOSITE DESIGN.................134
CHAPTER 8 EXERCISES...138

CHAPTER 9 COMPOSITE FIRST PLY FAILURE CRITERIA........**143**

9.1 INTRODUCTION TO STRUCTURAL FAILURE 143
Empirical Data 144
Empirical Models 144
Hypothesis 144
Theory 144
Physics 144
9.2 TEST SPECIMENS FOR STIFFNESS AND STRENGTH PROPERTIES 146
0° Tension Test Specimen 147
0° Compression Test Specimen 148
90° Tension Test Specimen 149
90° Compression Test Specimen 150
+/−45 Tension Test Specimen 151
9.3 EMPIRICAL CURVE-FIT FIRST PLY FAILURE CRITERIA 152
Maximum Stress First Ply Failure Criteria 152
Maximum Strain First Ply Failure Criteria 153
Tsai-Hill First Ply Failure Criteria 154
Tsai-Wu First Ply Failure Criteria 157
CHAPTER 9 EXERCISES 159

CHAPTER 10 COMPOSITE ANALYSIS **161**

10.1 COMPOSITE ZONE-BASED SHELL MODELING 163
10.2 COMPOSITE PLY-BASED SHELL MODELING 167
10.3 COMPOSITE PLY-BY-PLY SOLID MODELING 172
CHAPTER 10 EXERCISES 175

CHAPTER 11 COMPOSITE DESIGN OPTIMIZATION **179**

11.1 COMPOSITE FREE-SIZE OPTIMIZATION 182
11.2 COMPOSITE SIZE OPTIMIZATION 189
11.3 COMPOSITE SHUFFLING OPTIMIZATION 192
11.4 FINAL DESIGN VERIFICATION 195
CHAPTER 11 EXERCISES 196

APPENDIX A OPTISTRUCT I/O OPTIONS REFERENCE **197**

CSTRAIN 197
CSTRESS 198
DISPLACEMENT 199
OUTPUT 200
STRAIN 201
STRESS 202
THICKNESS 203
TITLE 204

APPENDIX B OPTISTRUCT SUBCASE CONTROL REFERENCE 205

ANALYSIS 205

DESGLB..*206*
DESOBJ...*207*
DESSUB...*208*
LABEL..*209*
LOAD...*210*
SPC...*211*
SUBCASE..*212*
TEMPERATURE...*213*

APPENDIX C OPTISTRUCT ANALYSIS BULK DATA REFERENCE

....**215**

CTRIA3...*215*
CQUAD4..*215*
CPENTA...*218*
CHEXA...*218*
CORD1C / CORD2C...*219*
CORD1R / CORD2R...*219*
DRAPE..*220*
FORCE..*221*
GRID...*222*
MAT1..*223*
MAT2..*224*
MAT8..*225*
MAT9..*227*
MAT9ORT..*228*
MOMENT...*229*
PCOMP...*230*
PCOMPG..*230*
PCOMPP...*230*
PLOAD2...*234*
PLOAD4...*235*
PLY..*236*
PSHELL..*238*
PSOLID...*240*
SET..*241*
SPC...*243*
STACK – Ply laminate definition...*244*
STACK - Interface laminate definition...*246*

APPENDIX D OPTISTRUCT OPTIMIZATION BULK DATA REFERENCE

....**249**

DCOMP..*249*
DCONADD..*257*
DCONSTR...*258*

DDVAL *259*
DESVAR *260*
DRESP1 .... *261*
DSHUFFLE *265*
DSIZE *267*
DVPREL1 .... *277*

APPENDIX E OPTISTRUCT COMPOSITE ANALYSIS MODELS. 279

OptiStruct Composite Zone-Based Shell Model.......................... 279
OptiStruct Composite Ply-Based Shell Model 281
OptiStruct Composite Ply-by-Ply Solid Model 283

APPENDIX F OPTISTRUCT COMPOSITE OPTIMIZATION MODELS 287

OptiStruct Composite Free-Size Model................................ 287
OptiStruct Composite Size Model 289
OptiStruct Composite Shuffling Model 292
OptiStruct Composite Final Design Model 294

APPENDIX G ANSWERS TO CHAPTER EXERCISES 297

Chapter 2 Exercise Answers 297
Chapter 3 Exercise Answers 299
Chapter 4 Exercise Answers 302
Chapter 5 Exercise Answers 308
Chapter 6 Exercise Answers 309
Chapter 7 Exercise Answers 314
Chapter 8 Exercise Answers 317
Chapter 9 Exercise Answers 321
Chapter 10 Exercise Answers 321
Chapter 11 Exercise Answers 321

REFERENCES 323

List of Figures

Figure 1.1, Particulate Composite Ply Material 1
Figure 1.2, Unidirectional Ply Material.. 2
Table 1.1, Typical Unidirectional Ply Material Properties (psi) 3
Table 1.2, Typical Engineering Metal Material Properties (psi).................... 3
Figure 1.3, 2D Plain Weave Ply Material... 4
Figure 1.4, 2D 5HS Weave Ply Material... 4
Figure 1.5, Honeycomb and Foam Core-Stiffened Composite Laminates...... 5
Figure 1.6, Ply Conventions 7
Figure 1.7, Laminated Plate Stacking Sequence and Z-coordinates 8
Figure 2.1, Coordinate System Definitions 12
Figure 2.2, General Body in Equilibrium 13
Figure 2.3, Stress Cube.. 13
Figure 2.4, Equilibrium of an Infinitesimal Element within a Body 18
Figure 2.5, Deformation of an Infinitesimal Element within a Body 21
Figure 2.6, Strain Cube.. 23
Figure 2.7, Definition of Direction Cosines for Transformation Laws 32
Figure 3.1, Boundary Condition #1 39
Figure 3.2, Boundary Condition #2 40
Figure 3.3, Boundary Condition #3 41
Figure 3.4, Boundary Condition #4 42
Figure 3.5, Boundary Condition #5 43
Figure 3.6, Boundary Condition #6 44
Figure 3.7, Transversely Isotropic Unidirectional Fiber-Matrix Composite. 52
Figure 3.8, Bar under Free Thermal Expansion 64
Figure 4.1, Representative Volume Element... 70
Figure 4.2, Representative Volume Element for Determination of E_1 71
Figure 4.3, Representative Volume Element for Determination of E_2 72
Figure 4.4, Representative Volume Element for Determination of υ_{12} 74
Figure 4.5, Representative Volume Element for Determination of υ_{23} 75
Figure 4.6, Representative Volume Element for Determination of G_{12} 77
Figure 4.7, Representative Volume Element for Determination of α_1 79
Figure 5.1, Plane Stress in a Single Ply... 93
Figure 5.2, Qbar Components vs. Theta.. 97
Figure 6.1, Laminated Plate Stacking Sequence and Z-coordinates 100
Figure 6.2, Kirchhoff Plate Definitions 101
Figure 6.3, Homogeneous Plate Resultant Force Sign Convention............. 102
Figure 6.4, Laminated Plate Resultant Force Sign Convention................... 103
Figure 6.5, Homogeneous Plate Resultant Moment Sign Convention 103
Figure 6.6, Laminated Plate Resultant Moment Sign Convention 104
Figure 6.7, Coupling Terms of the ABD Matrix 107
Figure 7.1, Equivalent Homogeneous Plate 113

Figure 7.2, Homogeneous Plate Stacking Sequence and Z-coordinate Definitions... 114
Figure 8.1, Plane Stress Stiffness Matrix in Global Coordinates vs. θ 131
Figure 8.2, Laminate with Unit Thickness Plies..................................... 131
Table 8.1, $(z_k^n - z_{k-1}^n)$ Geometric Terms for Unit Thick Plies.................. 132
Figure 8.3, Core-Stiffened Laminate Stacking Sequence and Z-Coordinates 135
Figure 9.1, 0° Tension Test Specimen 147
Figure 9.2, Typical 0° Tension Test Load vs. Strain................................. 147
Figure 9.3, 0° Compression Test Specimen 148
Figure 9.4, Typical 0° Compression Test Load vs. Strain.......................... 148
Figure 9.5, 90° Tension Test Specimen 149
Figure 9.6, Typical 90° Tension Test Load vs. Strain............................... 149
Figure 9.7, 90° Compression Test Specimen 150
Figure 9.8, Typical 90° Compression Test Load vs. Strain......................... 150
Figure 9.9, 45/−45 Tension Specimen 151
Figure 9.10, Typical 45/−45 Tension Text Load vs. Strain 151
Figure 9.11, Empirical Curve-Fit First Ply Failure Envelopes 158
Figure 10.1, Composite Plate Analysis Model.. 162
Figure 10.2, Zone-Based Element Normal and Stacking Sequence Definitions 163
Figure 10.3, Zone-Based Element Material System and Ply Orientation ... 164
Figure 10.4, Ply-Based Element Normal and Stacking Sequence Definitions 167
Figure 10.5, Ply-Based Element Material System and Ply Orientation 168
Figure 10.6, Solid Element Material Coordinate System Definition 172
Figure 11.1, Open Hole Tension Design Optimization Problem 181
Figure 11.2, Open Hole Tension Model 181
Figure 11.3, Composite Free-Size Design Variables............................... 183
Figure 11.4, Total Laminate Thickness Manufacturing Constraint 184
Figure 11.5, Ply Group Percentage Manufacturing Constraint.................. 184
Figure 11.6, Ply Group Balancing Manufacturing Constraint 185
Figure 11.7, Ply Group Constant Thickness Constraint............................ 185
Figure 11.8, Ply Group Drop Off Constraint 185
Figure 11.9, Composite Free-Size Ply Shape Generation......................... 187
Figure 11.10, Composite Free-Size 0° Ply Shapes................................ 188
Figure 11.11, Composite Free-Size 90° Ply Shapes.............................. 188
Figure 11.12, Composite Free-Size 45°/-45° Ply Shapes........................... 189
Figure 11.13, Composite Size Optimization Ply Shapes 192
Table 11.1, Composite Size Optimization Results – Number of Plies 192
Table 11.2, Composite Shuffle Optimization Results – Final Stacking Sequence 194
Figure 11.14, Fiber Direction Mechanical Strain at Design Load 195
Figure 11.15, First Ply Failure Index at Design Load................................ 195

Introduction

The main objective for writing this book was to provide professional engineers with a practical resource on the design and analysis of laminated composite structures. With the recent high utilization of composite materials in engineered structures—including aerospace, automotive, civil, and marine structures—comes the high demand for engineers with composite design and analysis knowledge and experience. However, the availability of engineers with significant experience in this field is difficult to obtain. Therefore, many engineers are faced with the daunting task of performing composite design and analysis projects with very little background in composites design and analysis. The book is aimed at helping those engineers gain practical composites design and analysis knowledge in as short a time as possible, and focuses on obtaining a practical understanding of the basic equations of composite material behavior as opposed to mathematical derivations of the same.

The book is written as a self-paced training course in which each chapter's material and exercises build on the previous chapters. It is recommended that readers progress through their study of the material in the sequence that follows. Chapter 1, Introduction to Composite Materials, introduces composites material terminology and conventions. Chapter 2, Elasticity Review, is a high-level review of basic equations of elasticity and strength of materials. It is important to review the material of chapter 2, as the mathematical conventions used throughout the book are introduced within this chapter. Chapter 3, Linear Elastic Material Laws, provides an introduction to the most common linear elastic material laws used to model composite ply materials. In addition, chapter 3 includes a detailed discussion on the importance of adding the effects of thermal strains into the material laws for composite ply materials. Chapter 4, Micromechanics of Composite Materials, introduces the equations and concepts of calculating homogenized composite ply material properties given the constituent material properties of the matrix and fiber that make up the composite ply material. Chapter 5, Macromechanical Behavior of a Ply, describes the behavior of a single ply material under in-plane loads as the principal material direction varies relative to the in-plane loading direction from $-90°$ to $90°$. Chapter 6, Classical Lamination Theory, builds on the behavior of a single ply material by describing the behavior of laminated plates, made by stacking multiple ply materials on top of each other, under in-plane load. Chapter 7, Homogenized Laminate Calculations, introduces the subject of calculating homogenized laminate properties given the [A], [B], and [D] matrices of a laminate— effectively modeling the laminate as if it were a single composite "thick ply" of material. Chapter 8, SMEAR Technology, builds upon the discussion of

Homogenized Laminate Calculations and stacking sequence effects to describe SMEAR technology and its application within the concept and early sizing design phases of composite structures design. Chapter 9, Composite First Ply Failure Criteria, discusses empirical and physics-based criteria that attempt to capture first ply failure events. Chapter 10, Solver Interfacing for Composite Analysis, applies the knowledge of chapters 1 through 9 to developing composite models for analysis with common industry standard solvers. Chapter 11, Composite Design and Optimization, expands upon chapter 10 to introduce composite design and optimization methods with common industry standard solvers. Appendices A, B, and C provide composite card references for OptiStruct (bulk data format) model building. Appendix D contains example OptiStruct analysis models. Appendix F contains example OptiStruct optimizaitn models. Finally appendix G contains the answes to the end of chapter exercise problems.

Chapter 1
Introduction to Composite Materials

The objective of this chapter is to present an overview of composite material terminology commonly used in practice. The chapter begins with a brief review of composite material product forms that are commonly used and generally available for designing of composite parts. One of the most broadly used composite material product forms used in highly-engineered products are fiber-matrix laminated composites, which are the principal subject of this book. The chapter continues by defining common elasticity and mechanics of materials terminology. Finally, the chapter concludes by defining ply and laminate conventions. All definitions and conventions defined in this chapter are used consistently throughout the book. Therefore, special attention to the contents of this chapter is recommended for further understanding of the remaining chapters.

1.1 Composite Materials

A composite material is a material that is formed by combining two or more materials on a macroscopic scale. The macroscopic scale is an important part of the definition of a composite material, as this book does not cover the analysis or design of composite materials at the microscopic scale. There are several basic composite material forms commonly used and currently available for producing composites parts, each of which is defined below.

Particulate Composites

Particulate composite materials are materials that are manufactured by spreading pieces of chopped fiber material onto a film of matrix material. This book does not cover the analysis or design of particulate composites.

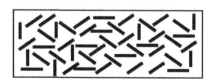

Figure 1.1, Particulate Composite Ply Material

Laminated Composites

Laminated composite materials are materials that are made up of any number of layered materials, of the same or different orientation, bonded together with a matrix material. The layers of a laminated composite, typically called plies, can be made from several materials, including adhesive plies, metallic-foil plies, fiber-matrix plies of various fiber and matrix material combinations, and core plies of various core materials.

Fiber-Matrix Laminated Composites

Fiber-matrix laminated composite materials are materials that are made up of any number of fiber-matrix plies, of the same or different fiber orientation, layered and bonded together with some matrix material. There are several types of fiber-matrix ply materials commonly used and currently available in the manufacturing of fiber-matrix laminated composite parts. The most common fiber materials used in fiber-matrix ply material systems include various types of boron, carbon, glass, and Kevlar® fibers. The most common matrix materials used in fiber-matrix ply material systems include various types of thermosetting and thermoplastic epoxies. Fiber-matrix ply materials can be classified into the following general categories; unidirectional plies, 2D plain weave plies, 2D 5-harness-satin (5HS) plies, and 3D woven fiber-matrix materials.

Unidirectional ply materials are made up of "straight" fibers embedded in a matrix material. Figure 1.2 shows a schematic of a unidirectional ply material. Typical unidirectional ply thicknesses range from 0.005–0.015 inches with typical fiber volumes between 0.45–0.70. Typical unidirectional ply material properties for several common fiber-matrix ply material systems are given in table 1.1. For comparative purposes, table 1.2 gives typical material properties for several common engineering metals. The properties presented in both tables should only be used within the exercises of this book and for general comparison purposes. The material property values presented in table 1.1 and table 1.2 should not be used for any actual design purposes.

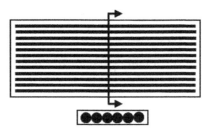

Figure 1.2, Unidirectional Ply Material

Table 1.1, Typical Unidirectional Ply Material Properties (psi)

Property (psi)	Boron-Epoxy	Carbon-Epoxy	Glass-Epoxy	Kevlar®-Epoxy
E_1	30.0e6	22.0e6	7.0e6	11.0e6
E_2	2.90e6	1.30e6	2.0e6	0.80e6
ν_{12}	0.26	0.30	0.25	0.34
ν_{23}	n/a	0.26	n/a	n/a
G_{12}	0.90e6	0.75e6	1.0e6	0.30e6
α_1 (ε/°F)	3.0e−6	−0.30e−6	3.0e−6	−2.0e−6
α_2 (ε/°F)	16.0e−6	18.0e−6	12.0e−6	40.0e−6
ρ (lbs/in^3)	0.072	0.056	0.065	0.052
X_t	190,000	170,000	150,000	200,000
X_c	380,000	170,000	150,000	38,000
Y_t	10,000	6,500	4,500	3,000
Y_c	35,000	28,000	18,000	15,000
S	14,000	10,000	8,000	6,000

Note: Properties are typical properties only and should not be used for design purposes.

Table 1.2, Typical Engineering Metal Material Properties (psi)

Property (psi)	Aluminum	Steel	Titanium
E	10.0e6	29.0e6	16.0e6
ν	0.33	0.30	0.31
G	3.76e6	11.2e6	6.11e6
α (/°F)	12.0e−6	6.0e−6	5.0e−6
ρ (lbs/in^3)	0.101	0.282	0.160
F_{tu}	60,000	125,000	130,000
F_{ty}	45,000	100,000	120,000

Note: Properties are typical properties only and should not be used for design purposes.

2D plain weave ply materials are made up of two unidirectional ply fibers woven into each other with a "1-over-1-under" pattern embedded in a matrix material. Typical 2D plain weave ply thicknesses range from 0.01–0.015 inches with typical fiber volumes between 0.04–0.065. 2D plain weave plies have reduced longitudinal stiffness as compared to their equivalent unidirectional ply product forms. The reduced stiffness is due to the undulation of the fibers that must be pulled taut, even though they are embedded within a matrix material, before complete load-carrying capacity can be achieved. Figure 1.3 shows a schematic of a 2D plain weave ply material.

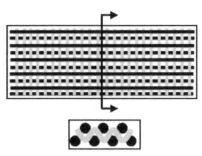

Figure 1.3, 2D Plain Weave Ply Material

2D 5-harness-satin (5HS) weave materials are made up of two unidirectional ply fibers woven into each other with a "1-under-4-over" pattern embedded in a matrix material. Typical 2D 5HS weave ply thicknesses range from 0.01–0.015 inches with typical fiber volumes between 0.04–0.06. 2D 5HS weave plies are longitudinally stiffer than their 2D plain weave ply counterparts. However, 2D 5HS weave plies still exhibit a decrease in longitudinal stiffness as compared to their equivalent unidirectional ply material. As can be seen from figure 1.4, which depicts a 2D 5HS weave ply material, approximately 80% of the 2D 5HS weave ply is equivalent to two orthogonal unidirectional plies. The remaining 20% of the 2D 5HS ply contains the undulations of the fibers due to the weave pattern. Therefore, typical 2D 5HS weave plies exhibit approximately 1% decrease in longitudinal stiffness as compared to their equivalent unidirectional ply material. Furthermore, it is appropriate to model a 2D 5HS weave ply material as two orthogonal unidirectional plies with the unidirectional longitudinal stiffness, E_1, reduced by 1% and the thickness of the unidirectional plies equal to ½ the total thickness of the 2D 5HS weave ply material.

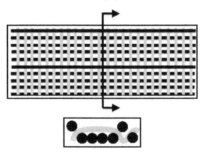

Figure 1.4, 2D 5HS Weave Ply Material

3D weaves are made up of unidirectional fibers woven in 3-dimensional patterns embedded within a matrix material. 3D weaves come in many different thickness and fiber volumes, and they are typically custom-made for specific purposes.

Core-Stiffened Laminated Composites

Core-stiffened composite laminates are made up of two laminated composite face sheets separated by a core material. Typical core materials include aluminum honeycomb, carbon honeycomb, and foam. The laminated composite face sheets can be any of the laminated composite material types defined above. Figure 1.5 shows a schematic of a honeycomb core-stiffened and a foam core-stiffened laminate composite.

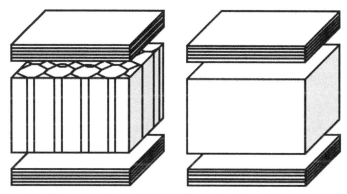

Figure 1.5, Honeycomb and Foam Core-Stiffened Composite Laminates

1.2 Material Terminology

Microscopic

Microscopic is a term used to describe physical objects smaller than can easily be seen by the naked eye and which require a lens or microscope to see clearly.

Macroscopic

Macroscopic is a term used to describe physical objects that are measurable and observable by the naked eye and do not require a lens or microscope to see clearly.

Micromechanics

Micromechanics is the study of composite material behavior wherein the interaction of the constituent materials is examined in detail as part of the definition and behavior of the heterogeneous composite material.

Macromechanics

Macromechanics is the study of composite material behavior wherein the material is assumed homogeneous and the effects of the constituent materials are detected only as averaged apparent properties, otherwise called effective properties, of the composite material.

Homogeneous

A homogeneous body has uniform properties throughout; thus, the material properties of the body are independent on the position within the body.

Heterogeneous

A heterogeneous body has non-uniform properties throughout; thus, the material properties of the body are dependent on the position within the body.

Isotropic

An isotropic body has material properties that are the same in all directions at a given point within a body; thus, the material properties are independent of orientation at a specified point within the body.

Orthotropic

An orthotropic body has material properties that are the same in each of three orthogonal planes at a given point within a body; thus, the material properties are dependent on orientation at a specified point within the body.

Anisotropic

An anisotropic body has material properties that are different in all directions at a given point within a body; thus, the material properties are dependent on orientation at a specified point within the body.

1.3 Ply Conventions

In the case of fiber-matrix unidirectional composite plies, the material coordinate system 1-axis defines the fiber direction, the material coordinate system 2-axis defines the transverse matrix direction, and the material coordinate system 3-axis defines the through-thickness direction of a ply. The fiber orientation of a ply, theta, is defined relative to the global system x-axis using right-hand rule to define positive theta as shown in figure 1.6.

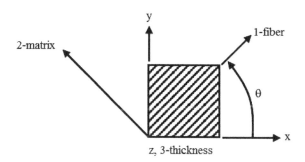

Figure 1.6, Ply Conventions

1.4 Laminate Conventions and Definitions

To facilitate discussions on laminated composites, we must first define the laminate stacking sequence and ply z-coordinate conventions that will be used throughout this book. The laminate stacking sequence and ply z-coordinate conventions used in this book are typical of the most popular FEA solvers. However, the conventions are not necessarily consistent with those used in all FEA solvers. It is recommended that users consult the specific FEA solver documentation for the laminate conventions used within that solver. Chapter 10 gives the relevant laminate stacking sequence and ply z-coordinate conventions for several popular solvers.

The global coordinate system of a homogeneous or laminated plate is defined in figure 1.7. Note that the xy-plane defined by the global coordinate system goes through the middle surface of the plate with the z-axis "down" as defined using right-hand rule for the xy-plane as shown in figure 1.7. Assuming a laminated plate, the laminate stacking sequence and ply z-coordinate conventions are also defined in figure 1.7. Plies are numbered 1 through n with the 1st ply defined as the most negative z ply and the nth ply as the most positive z ply. Plies stack in the positive z-axis direction (plate normal) from ply 1 to ply n. In addition, the z-coordinate value for the kth ply is always defined as the most positive z-coordinate interface for that ply. The z-coordinate of a ply is measured relative to the middle surface of the plate.

Therefore, ply 1 through ply $(n / 2) - 1$ will have negative z-coordinates. Likewise, ply $(n / 2)$ to ply n will have positive z-coordinates. Since, by definition, there will always be $(n + 1)$ ply interfaces, the z_0 coordinate is typically defined as $-t/2$ unless a laminate offset is applied.

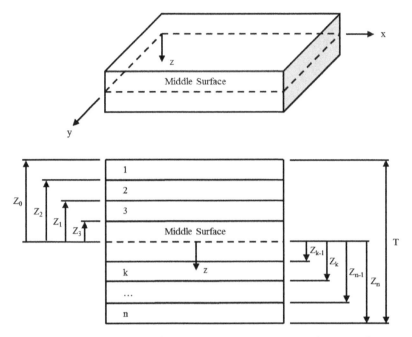

Figure 1.7, Laminated Plate Stacking Sequence and Z-coordinates

Defining Laminates

Laminates are typically specified in the engineering community using the following notation:

$$[\text{ply1} / \text{ply2} / (\text{ply3} / \text{ply4})_n / \dots / \underline{\text{ply n}}]_{n\ s}$$

The subscript n defines the number of repeating units within its given brackets, and the subscript s defines a laminate as a symmetric laminate. For symmetric laminates, only the negative z-coordinate plies are specified, as the positive z-coordinate plies of the laminate can be readily determined from the symmetric definition. In addition, an underlined ply specifies the ply as being symmetric about ½ of the ply and is not repeated on the other half of the laminate. An underlined ply allows for symmetric laminate definitions containing an odd number of plies; otherwise, symmetric laminates will always have an even number of plies.

Symmetric Laminates

A symmetric laminate is defined as a laminate that is composed of plies such that the thickness, angle (theta), and material of the plies are symmetric about the middle surface of the laminate. For symmetric laminates, the [B] matrix is zero and exhibits no extensional–bending or shear–twisting coupling behaviors. Examples of symmetric laminates using engineering notation, assuming all plies have the same thickness and material, are given below.

[0/45/90/45/0], which can also be written as [0/45/90]$_s$
[45/-45/90/0/0/90/−45/45], which can also be written as [45/−45/90/0]$_s$

Anti-Symmetric Laminates

An anti-symmetric laminate is defined as a laminate for which every +θ ply and −θ ply on the negative z-half of the laminate, there exist a −θ ply and +θ ply respectively on the positive z-half of the laminate with the same thickness and material at the same stacking sequence location. In addition, 0 plies and 90 plies must be symmetric about the middle surface of the laminate. Examples of anti-symmetric laminates using engineering notation, assuming all plies have the same thickness and material, are given below.

[0/90/−45/45/90/0]
[0/45/90/−45/45/90/−45/0]

Balanced Laminates

A balanced laminate is defined as a laminate for which every +θ ply there exists a −θ ply of the same thickness and material. The definition of a balanced laminate does not define where in the laminate stacking sequence the plies exist, just that there are the same number of +θ plies and −θ plies in total for the laminate. Balanced laminates have zero A_{14} and A_{24} components and exhibit no extensional–shear coupling behavior. In addition, if a balanced laminate is also anti-symmetric, then the laminate will additionally have zero D_{14} and D_{24} components and will also not exhibit bending-twisting coupling behavior. Examples of balanced laminates using engineering notation, assuming all plies have the same thickness and material, are given below.

[45/−45/−30/30]
[22.5/−22.5/90/−22.5/22.5]

Cross-Ply Laminates

A cross-ply laminate is defined as a laminate composed of only 0 plies and 90 plies of the same thickness and material. Cross-ply laminates have zero A_{14}, A_{24}, D_{14}, and D_{24} components and exhibit no extensional-shear or bending-twisting coupling behaviors. In addition, if a cross-ply laminate is symmetric, then the laminate will additionally have a zero [B] matrix and exhibit no extensional-bending or shear-twisting coupling behaviors. Examples of cross-ply laminates using engineering notation, assuming all plies have the same thickness and material, are given below.

[0/90/0/90/0]
[0/0/90/90/0/0/90]$_s$

Angle-Ply Laminates

An angle-ply laminate is defined as a laminate composed of only $+\theta$ plies and $-\theta$ plies of the same thickness and material. In general, angle-ply laminates have fully populated [A], [B], and [D] matrices. However, if an angle-ply laminate is balanced, then the laminate will have zero A_{14} and A_{24} components and exhibit no extensional-shear coupling behavior. In addition, if an angle-ply laminate is symmetric, then the laminate will additionally have a zero [B] matrix and exhibit no extensional-bending or shear-twisting coupling behaviors. Examples of angle-ply laminates using engineering notation, assuming all plies have the same thickness and material, are given below.

[45/−45/−30/30]
[−30/30/60/30/−30]

General Laminates

A general laminate is defined as a laminate that does not fall into any of the previous laminate definitions. General laminates generally exhibit fully populated [A], [B], and [D] matrices, and therefore all types of coupling typically exist; extension-shear coupling (A_{14} and A_{24} terms), extension-bending coupling ([B] matrix terms), shear-twisting coupling ([B] matrix terms), bending-twisting coupling (D_{14} and D_{24} terms

Examples of general laminates using engineering notation, assuming all plies have the same thickness and material, are given below.

[0/45/90/22.5/0/45]
[90/−45/0/90/−45/0]

Chapter 2
Elasticity Review

This chapter reviews elasticity conventions that will be used throughout the book. There are two approaches to solving problems of determining deformation, strain, and stress within a body; elasticity and strength of materials. These two approaches differ largely in the assumptions made of the displacement field within a body. The strength of materials approach makes assumptions about the displacement field within a body for specific conditions of practical problems encountered in engineering. These displacement field assumptions simplify the resulting calculations for stress and strain within a body. The equations derived using the strength of materials approach are valid only if the specific conditions under which the displacement field was assumed are valid. Elasticity, on the other hand, does not make as many assumptions about the displacement field within a body and attempts to solve for the "exact" displacement, strain, and stress fields within a body. Equations derived using the theory of elasticity are more general in application but require significantly more mathematical analysis. Classical lamination theory, which is the major topic of discussion within this book, implements the strength of materials approach to obtaining the displacement, strain, and stress fields within a laminated plate under various conditions of practical problems encountered in engineering. However, it is still important to understand the basic equations of elasticity so that the simplifying assumptions used to derive the equations of classical lamination theory can be understood and verified.

2.1 Coordinate Systems

Within this book, the (xyz) coordinate system refers to a general coordinate system or the global coordinate system depending on context. The (XYZ) coordinate system always refers to the global coordinate system. Finally, the (123) coordinate system always refers to the material coordinate system. For a fiber-matrix unidirectional composite ply, the material coordinate system 1-axis is defined as the fiber direction, the material coordinate system 2-axis is defined as the transverse matrix direction, and the material coordinate system 3-axis is defined as the through-thickness direction of the ply. In this chapter, we will reference all equations of elasticity in the (xyz) general coordinate system. However, the equations of elasticity are equally applicable in either the (XYZ) global coordinate system or in the (123) material coordinate system with appropriate transformations. Figure 2.1 details the various coordinate systems that are utilized within this book.

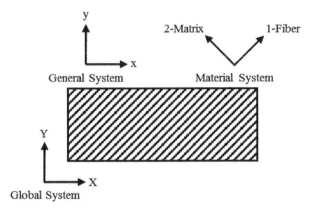

Figure 2.1, Coordinate System Definitions

2.2 Stress

In order to define stress, we shall consider a general body in equilibrium subject to a system of external forces as shown in figure 2.2. If the body is cut at a plane through point Q, an internal force distribution must exist within the plane to keep the body in equilibrium. In general, the internal force varies from point to point within the cut plane. The section of area, ΔA, within the cut plane at point Q is acted on by a force, ΔF, which in general does not lie along any of the x-, y-, or z-axis directions. Furthermore, decomposing ΔF into its components parallel to the x-, y-, and z-axis directions (F_x, F_y, and F_z respectively) and taking the limit as ΔA goes to zero at point Q, we define the normal stress, σ_x, and shearing stresses, τ_{xy} and τ_{xz}, within the plane at point Q as given by equation (2.1). Furthermore, it should be obvious that the normal stress and shearing stresses vary from point to point on the plane, as ΔF varies from point to point on the plane. However, stress does not just depend on ΔF; it also depends on the orientation of the plane through a given point. Take, for instance, another plane through point Q with a different orientation than previously discussed. This new plane section will produce a different internal resultant force at point Q and hence different normal stress and shearing stresses on the new plane at point Q. The complete description of stress at a point therefore requires knowledge of the stress on all planes passing through the point. It turns out that in order to determine the stress on an infinite number of planes passing through a point it is only required to specify the stress on three mutually perpendicular planes passing through a point. These three mutually perpendicular planes at a point can be thought of as forming an infinitesimal cube at the point, called the stress cube, as is depicted in figure 2.3.

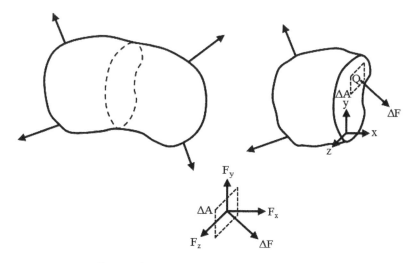

Figure 2.2, General Body in Equilibrium

$$\sigma_x = \lim_{\Delta A \to 0} \frac{\Delta F_x}{\Delta A} = \frac{dF_x}{dA} \ , \ \tau_{xy} = \lim_{\Delta A \to 0} \frac{\Delta F_y}{\Delta A} = \frac{dF_y}{dA} \ , \ \tau_{xz} = \lim_{\Delta A \to 0} \frac{\Delta F_z}{\Delta A} = \frac{dF_z}{dA}$$

$$(2.1)$$

Stress is a 2nd order tensor requiring two indexes to identify its components. The first index indicates the normal direction of the plane on which the stress component acts. The second index indicates the direction of the stress component on that plane. For example, τ_{xy} is the shearing stress on the plane that has the x-axis direction normal to it and acts in the y-axis direction on this plane, as shown in figure 2.3.

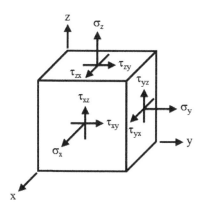

Figure 2.3, Stress Cube

Stress tensors can be represented several different ways.   The tensor representation for stress is given in equation (2.2). Notice that the stress tensor is symmetric; thus, six stress components completely define the 3D state of stress within a given plane at a given point within a body. Note that there are an infinite number of planes on which a stress tensor can act at a given point within a body. Therefore, a stress tensor is meaningless unless the plane on which it acts is specified. The shorthand notation used for defining the plane on which the stress tensor acts is a single subscript, outside the brackets, which identifies the plane. As shown in equation (2.2), the x subscript outside the brackets identifies a stress tensor as acting in a general rectangular coordinate system. Additionally, a 1 subscript outside the brackets would identify the stress tensor as acting in the principal material coordinate system.   If no subscript is given for a stress tensor written in shorthand notation, then the plane on which the stress tensor acts should be able to be determined from the context of the discussion or does not add clarity to a specific derivation.

$$\{\sigma\}_x = \begin{bmatrix} \sigma_{xx} & \tau_{xy} & \tau_{xz} \\ \tau_{xy} & \sigma_{yy} & \tau_{yz} \\ \tau_{xz} & \tau_{yz} & \sigma_{zz} \end{bmatrix} \qquad (2.2)$$

The tensor representation for stress is also commonly written in a contracted vector representation, as shown in equation (2.3). The contracted vector representation will be used most commonly throughout this book.

$$\{\sigma\}_x = \begin{Bmatrix} \sigma_x \\ \sigma_y \\ \sigma_z \\ \tau_{xy} \\ \tau_{yz} \\ \tau_{xz} \end{Bmatrix} \qquad (2.3)$$

2.3 Stress Invariants

The principal stresses, σ_1, σ_2, and σ_3 for a given stress tensor, are the stresses that act on three mutually orthogonal planes that have zero shear stress. The principal stresses can be determined by solving equation (2.4), where σ_p denotes one of the three principal stresses, and l, m, and n denote the direction cosines for the orientation of the principal plane on which the given principal stress acts relative to the orientation of the given stress tensor.

$$\begin{bmatrix} \sigma_x - \sigma_p & \tau_{xy} & \tau_{xz} \\ \tau_{xy} & \sigma_y - \sigma_p & \tau_{yz} \\ \tau_{xz} & \tau_{yz} & \sigma_z - \sigma_p \end{bmatrix} \begin{Bmatrix} l \\ m \\ n \end{Bmatrix} = \begin{Bmatrix} 0 \\ 0 \\ 0 \end{Bmatrix} \tag{2.4}$$

In order for equation (2.4) to have a non-trivial solution, the characteristic determinate of equation (2.4) must vanish. This can be written as equation (2.5).

$$\begin{vmatrix} \sigma_x - \sigma_p & \tau_{xy} & \tau_{xz} \\ \tau_{xy} & \sigma_x - \sigma_p & \tau_{yz} \\ \tau_{xz} & \tau_{yz} & \sigma_x - \sigma_p \end{vmatrix} = 0 \tag{2.5}$$

Expanding equation (2.5) leads to the stress cubic equation (2.6). The three roots of the stress cubic equation are the principal stresses; σ_1, σ_2, and σ_3. Since the principal stresses are independent of the original orientation of the given stress tensor; I_1, I_2, and I_3 must be independent of the original orientation. Thus; I_1, I_2, and I_3 are known as the stress invariants and are useful for validating stress tensor transformations. It should also be observed that the principal stresses are the eigenvalues of equation (2.4) and the principal stress directions are the eigenvectors of equation (2.4).

$$\sigma_p^3 - I_1\sigma_p^2 + I_2\sigma_p - I_3 = 0 \tag{2.6}$$

Where;

$$I_1 = \sigma_x + \sigma_y + \sigma_z$$

$$I_2 = \sigma_x\sigma_y + \sigma_x\sigma_z + \sigma_y\sigma_z - \tau_{xy}^2 - \tau_{yz}^2 - \tau_{xz}^2$$

$$I_3 = \begin{vmatrix} \sigma_x & \tau_{xy} & \tau_{xz} \\ \tau_{xy} & \sigma_y & \tau_{yz} \\ \tau_{xz} & \tau_{yz} & \sigma_z \end{vmatrix}$$

In general, the solution to the stress cubic equation is difficult to obtain. However, a practical approach to determining the roots of the stress cubic equation is given in Reference 4, Appendix B.

The stress cubic equation can be simplified in a 2D plane stress environment to obtain the 2D plane stress principal stresses, as given by equation (2.7).

$$\sigma_p^2 - I_1\sigma_p + I_2 = 0 \tag{2.7}$$

Where;

$$I_1 = \sigma_x + \sigma_y$$

$$I_2 = \sigma_x\sigma_y - \tau_{xy}^2$$

Solving equation (2.7) using the quadratic formula yields the 2D plane stress principal stresses, as given by equation (2.8).

$$\sigma_{1,2} = \frac{\sigma_x + \sigma_y}{2} \pm \sqrt{\left(\frac{\sigma_x - \sigma_y}{2}\right)^2 + \tau_{xy}^2} \tag{2.8}$$

The octahedral shear stress is the maximum shear stress that exists in a 3D state of stress. The octahedral shear stress acts on octahedral planes relative to the principal planes. However, unlike the principal stress planes that have zero shear stress, the octahedral shear stress planes have a normal stress component, called the octahedral normal stress or the mean stress. The octahedral shear stress and the corresponding mean stress, in terms of the principal stresses, are given by equations (2.9) and (2.10) respectively. The octahedral shear stress can also be written in terms of the stress tensor components, as given by equation (2.11)

$$\tau_{oct} = \frac{1}{3}\sqrt{(\sigma_1 - \sigma_2)^2 + (\sigma_2 - \sigma_3)^2 + (\sigma_3 - \sigma_1)^2} \tag{2.9}$$

$$\sigma_{mean} = \frac{1}{3}(\sigma_1 + \sigma_2 + \sigma_3) \tag{2.10}$$

$$\tau_{oct} = \frac{1}{3}\sqrt{(\sigma_x - \sigma_y)^2 + (\sigma_y - \sigma_z)^2 + (\sigma_z - \sigma_x)^2 + 6(\tau_{xy}^2 + \tau_{yz}^2 + \tau_{xz}^2)}$$

$$\tag{2.11}$$

The octahedral stresses can be simplified in a 2D plane stress environment to obtain the 2D plane stress maximum shear stress and mean stress, as given by equations (2.12) and (2.13) respectively.

$$\tau_{max} = \frac{1}{2}(\sigma_1 - \sigma_2) \tag{2.12}$$

$$\sigma_{mean} = \frac{1}{2}(\sigma_1 + \sigma_2) \tag{2.13}$$

The von Mises stress is associated with the maximum distortional energy yield theory, also known as the von Mises yield theory. The maximum distortional energy yield theory states that failure by yielding occurs within a body when; the distortional energy per unit volume in a 3D stress state at any point within the body equals the distortional energy per unit volume at yielding of a simple tension test. In order to dissect this statement into a mathematical relationship, we need to define the concept of strain energy. Strain energy is the work done by external forces in causing deformation of a body. The strain energy density U_o, or stain energy per unit volume, is the work done by internal stresses causing strain within a body, as given by equation (2.14). The strain energy density can be broken into two parts; dilatational strain energy density U_{ov}, and distortional strain energy density U_{od}, as given by equation (2.15). The dilatational strain energy density is the portion of the strain energy density which causes a change in volume with no change in shape. For an isotropic body, the dilatational strain energy density is given by equation (2.16). The distortional strain energy density is the portion of the strain energy density which causes a change in shape with no change in volume. For an isotropic body, the distortional strain energy density is given by equation (2.17).

$$U_o = \frac{1}{2}(\sigma_x \varepsilon_x + \sigma_y \varepsilon_y + \sigma_z \varepsilon_z + \tau_{xy} \gamma_{xy} + \tau_{yz} \gamma_{yz} + \tau_{xz} \gamma_{xz}) \qquad (2.14)$$

$$U_o = U_{ov} + U_{od} \qquad (2.15)$$

$$U_{ov} = \frac{3(1-2v)}{2E} \sigma_{mean}^2 \qquad (2.16)$$

$$U_{od} = \frac{3(1+v)}{2E} \tau_{oct}^2 \qquad (2.17)$$

The final piece to the maximum distortional energy yield theory is calculation of the distortional strain energy density at yielding of a simple tension test, as given by equation (2.18).

$$U_{od} = \frac{(1+v)}{3E} F_{ty}^2 \quad \text{at yielding from a simple tension test} \qquad (2.18)$$

Equating equations (2.17) and (2.18), the maximum distortional energy yield theory, also known as the von Mises yield theory, is derived. Failure by yielding will occur when the relationship given by equation (2.19) is satisfied.

$$\sqrt{\frac{9}{2}\tau_{oct}^2} = F_{ty} \quad \text{or} \quad \frac{9\tau_{oct}^2}{2F_{ty}^2} = 1 \qquad (2.19)$$

The left-hand side of equation (2.19), known as the von Mises stress σ_{vm}, can be rewritten in terms of the principal stresses as given by equation (2.20), or in terms of the stress tensor components as given by equation (2.20a).

$$\sigma_{vm} = \frac{1}{\sqrt{2}} \sqrt{(\sigma_1 - \sigma_2)^2 + (\sigma_2 - \sigma_3)^2 + (\sigma_3 - \sigma_1)^2} \qquad (2.20)$$

$$\sigma_{vm} = \frac{1}{\sqrt{2}} \sqrt{(\sigma_x - \sigma_y)^2 + (\sigma_y - \sigma_z)^2 + (\sigma_z - \sigma_x)^2 + 6(\tau_{xy}^2 + \tau_{yz}^2 + \tau_{xz}^2)}$$

$$(2.20a)$$

The von Mises Stress in 2D plane stress is given by equation (2.21).

$$\sigma_{vm} = \sqrt{\sigma_1^2 - \sigma_1\sigma_2 + \sigma_2^2} \qquad (2.21)$$

2.4 Equilibrium Equations

In order to derive the equilibrium equations, we will consider an infinitesimal element within the general body of figure 2.2 under plane stress conditions as shown in figure 2.4. Assuming that the stress on each face is uniform and that the element is of unit thickness we can sum of the forces and moments to zero in order to derive the equilibrium equations. The terms F_x and F_y are body forces that act over a volume of material; thus, they are forces per unit volume.

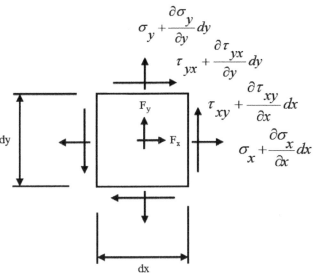

Figure 2.4, Equilibrium of an Infinitesimal Element within a Body

Summing the forces in the x-direction to zero we obtain;

$$\left(\sigma_x + \frac{\partial \sigma_x}{\partial x} dx\right) dy - \sigma_x dy + \left(\tau_{yx} + \frac{\partial \tau_{yx}}{\partial y} dy\right) dx - \tau_{yx} dx + F_x dxdy = 0$$

Which simplifies to;

$$\left(\frac{\partial \sigma_x}{\partial x} + \frac{\partial \tau_{yx}}{\partial y} + F_x\right) dxdy = 0$$

Furthermore, since dxdy cannot be zero, the quantity in parenthesis must vanish and consequently we obtain;

$$\frac{\partial \sigma_x}{\partial x} + \frac{\partial \tau_{yx}}{\partial y} + F_x = 0$$

Similarly, summing the forces in the y-direction to zero we obtain;

$$\frac{\partial \sigma_y}{\partial y} + \frac{\partial \tau_{xy}}{\partial x} + F_y = 0$$

Summing the moments about the lower left hand corner we obtain;

$$\sigma_x dy \frac{dy}{2} - \left[\left(\sigma_x + \frac{\partial \sigma_x}{\partial x} dx\right) dy\right] \frac{dy}{2} - \sigma_y dx \frac{dx}{2} + \left[\left(\sigma_y + \frac{\partial \sigma_y}{\partial y} dy\right) dx\right] \frac{dx}{2}$$

$$+ \left[\left(\tau_{xy} + \frac{\partial \tau_{xy}}{\partial x} dx\right) dy\right] dx - \left[\left(\tau_{yx} + \frac{\partial \tau_{yx}}{\partial y} dy\right) dx\right] dy$$

$$- F_x dxdy \frac{dy}{2} + F_y dxdy \frac{dx}{2} = 0$$

Which simplifies to;

$$\frac{\partial \sigma_x}{\partial x} \frac{dxdy^2}{2} + \frac{\partial \sigma_y}{\partial y} \frac{dx^2 dy}{2} + \tau_{xy} dxdy + \frac{\partial \tau_{xy}}{\partial x} \frac{dx^2 dy}{2} - \tau_{yx} dxdy + \frac{\partial \tau_{yx}}{\partial y} \frac{dxdy^2}{2}$$

$$- F_x \frac{dxdy^2}{2} + F_y \frac{dx^2 dy}{2} = 0$$

Ignoring the infinitely small triple product terms of dx^2dy and $dxdy^2$ we obtain;

$$\tau_{xy} = \tau_{yx}$$

Expanding the infinitesimal element of figure 2.4 to three dimensions and summing the forces in all three directions, the three-dimensional equilibrium equations are derived as shown in equation (2.22).

$$\frac{\partial \sigma_x}{\partial x} + \frac{\partial \tau_{xy}}{\partial y} + \frac{\partial \tau_{xz}}{\partial z} + F_x = 0$$

$$\frac{\partial \sigma_y}{\partial y} + \frac{\partial \tau_{xy}}{\partial x} + \frac{\partial \tau_{yz}}{\partial z} + F_y = 0 \qquad (2.22)$$

$$\frac{\partial \sigma_z}{\partial z} + \frac{\partial \tau_{xz}}{\partial x} + \frac{\partial \tau_{yz}}{\partial y} + F_z = 0$$

In addition, summing the moments in all three directions of the three-dimensional infinitesimal element we obtain the result showing that only six of the nine stress components at a point are independent.

$$\tau_{xy} = \tau_{yx} \qquad\qquad \tau_{yz} = \tau_{zy} \qquad\qquad \tau_{xz} = \tau_{zx} \qquad (2.23)$$

2.5 Strain

The components of displacement at a point within the general body of figure 2.2 occurring in the x-, y-, and z-axes directions are denoted by u, v, and w respectively. The displacement at every point within the general body constitutes the displacement field, as given by equations (2.24). Considering the general body to be under plane strain conditions, an infinitesimal element ABCD will displace to A'B'C'D' under this condition, as shown in figure 2.5. The displaced element A'B'C'D' exhibits both rigid body translation and deformation. The overall deformation consists of two distinct deformations; changes in length of each side of the element, and rotations of the sides of the element with respect to each other. The normal strain, ε, is defined as the unit change in length of each side, as given by equation (2.25). The normal strain is positive if the side increases in length and is negative if the side decreases in length. The shearing strain, γ, is defined as the change in angle between the sides of the element from its original right angle, as given by equation (2.28). The shear strain is measured in radians and is positive if the right angle between the sides in the positive axes directions decreases.

$$u = u(x, y, z)$$
$$v = v(x, y, z)$$
$$w = w(x, y, z)$$
(2.24)

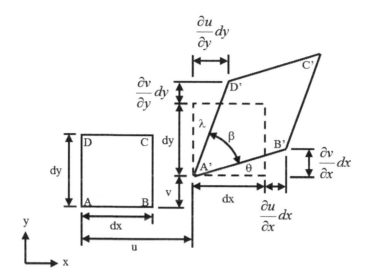

Figure 2.5, Deformation of an Infinitesimal Element within a Body

The normal strain is defined as;

$$\varepsilon_x = \frac{\Delta L}{L} = \frac{A'B'-AB}{AB} = \frac{A'B'-dx}{dx}$$
(2.25)

Ignoring higher order terms as negligible compared to the magnitude of dx, the length of A'B' can be approximated in the small deformation theory of strain as;

$$A'B' = dx + \frac{\partial u}{\partial x} dx$$
(2.26)

Substituting equation (2.15) into equation (2.14) the definition of normal strain in the x-direction is derived for small strain theory as given by equation (2.16).

$$\varepsilon_x = \frac{\partial u}{\partial x}$$
(2.27)

The shear strain is defined as;

$$\gamma_{xy} = \frac{\pi}{2} - \beta = \theta - \lambda \tag{2.28}$$

Using the equation $s = r\theta$ we can write θ and λ as;

$$\theta = \frac{\dfrac{\partial v}{\partial x} dx}{dx + \dfrac{\partial u}{\partial x} dx} \tag{2.29}$$

$$\lambda = \frac{-\dfrac{\partial u}{\partial y} dy}{dy + \dfrac{\partial v}{\partial y} dy} \tag{2.30}$$

Neglecting $\partial u / \partial x$ and $\partial u / \partial x$ as small compared to the magnitude of dx and dy respectively, and substituting equation (2.29) and (2.30) into equation (2.28), the definition of shear strain in the xy-direction is derived for small-strain theory as given by equation (2.31).

$$\gamma_{xy} = \frac{\partial u}{\partial y} + \frac{\partial v}{\partial x} \tag{2.31}$$

Expanding the infinitesimal element of figure 2.5 to three dimensions and applying similar normal strain and shear strain definitions, the three dimensional strain displacement equations are derived as shown in equation (2.32).

$$\varepsilon_x = \frac{\partial u}{\partial x}$$

$$\varepsilon_y = \frac{\partial v}{\partial y}$$

$$\varepsilon_x = \frac{\partial w}{\partial z} \tag{2.32}$$

$$\varepsilon_{xy} = \frac{1}{2}\left(\frac{\partial u}{\partial y} + \frac{\partial v}{\partial x} \right)$$

$$\varepsilon_{yz} = \frac{1}{2}\left(\frac{\partial v}{\partial z} + \frac{\partial w}{\partial y} \right)$$

$$\varepsilon_{xz} = \frac{1}{2}\left(\frac{\partial u}{\partial z} + \frac{\partial w}{\partial x} \right)$$

Strain is a 2nd order tensor requiring two indexes to identify its components. The first index indicates the direction of the normal to the plane on which the strain component acts. The second index indicates the direction of the strain component within the plane. For example, γ_{zy} is the engineering shearing strain on the plane that has the z-axis direction normal and acts in the y-axis direction on this plane, as shown in figure 2.6.

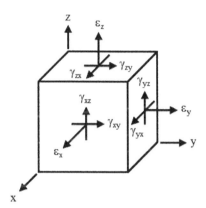

Figure 2.6, Strain Cube

Strain tensors can be represented several ways. The tensor representation for strain is given in equation (2.33). Notice that the strain tensor is symmetric; thus, six strain components completely define the 3D state of strain within a given plane at a point within a body. Note that there are an infinite number of planes on which a strain tensor can act at a given point within a body. Therefore, a strain tensor is meaningless unless the plane on which it acts is specified. The shorthand notation used for defining the plane on which the strain tensor acts is a single subscript, outside the brackets, which identifies the plane. As shown in equation (2.33), the x subscript outside the brackets identifies this strain tensor as acting in a general rectangular coordinate system. Additionally, a 1 subscript outside the brackets identifies the strain tensor as acting in the principal material coordinate system. If no subscript is given for a strain tensor written in shorthand notation, then the plane on which the strain tensor acts should be able to be determined from the context of the discussion or does not add clarity to a specific derivation.

$$\{\varepsilon\}_x = \begin{bmatrix} \varepsilon_{xx} & \varepsilon_{xy} & \varepsilon_{xz} \\ \varepsilon_{xy} & \varepsilon_{yy} & \varepsilon_{yz} \\ \varepsilon_{xz} & \varepsilon_{yz} & \varepsilon_{zz} \end{bmatrix} \qquad (2.33)$$

The tensor representation for strain is also commonly written in an engineering contracted vector representation, as shown in equation (2.34). The engineering contracted vector representation for strain will be used most commonly throughout this book. Note that the engineering shear strain definition is twice that of the tensor shear strain definition or the total angle change at a vertex, as shown in figure 2.5. Consequently, engineering shear strains need to be converted into tensor shear strains before any tensor transformation laws can be applied.

$$\{\varepsilon\}_x = \begin{Bmatrix} \varepsilon_{xx} \\ \varepsilon_{yy} \\ \varepsilon_{zz} \\ 2\varepsilon_{xy} \\ 2\varepsilon_{yz} \\ 2\varepsilon_{xz} \end{Bmatrix} = \begin{Bmatrix} \varepsilon_x \\ \varepsilon_y \\ \varepsilon_z \\ \gamma_{xy} \\ \gamma_{yz} \\ \gamma_{xz} \end{Bmatrix} \tag{2.34}$$

2.6 Strain Invariants

The principal strains, ε_1, ε_2, and ε_3 for a given strain tensor, are the strains that act on three mutually orthogonal planes that have zero shear strain. The principal strains can be determined by solving equation (2.35), where σ_p denotes one of the three principal stresses and l, m, and n denote the direction cosines for the orientation of the principal plane on which the given principal stress acts relative to the orientation of the given stress tensor.

$$\begin{bmatrix} \varepsilon_x - \varepsilon_p & \dfrac{\gamma_{xy}}{2} & \dfrac{\gamma_{xz}}{2} \\ \dfrac{\gamma_{xy}}{2} & \varepsilon_y - \varepsilon_p & \dfrac{\gamma_{yz}}{2} \\ \dfrac{\gamma_{xz}}{2} & \dfrac{\gamma_{yz}}{2} & \varepsilon_z - \varepsilon_p \end{bmatrix} \begin{Bmatrix} l \\ m \\ n \end{Bmatrix} = \begin{Bmatrix} 0 \\ 0 \\ 0 \end{Bmatrix} \tag{2.35}$$

In order for equation (2.35) to have a non-trivial solution, the characteristic determinate of equation (2.35) must vanish. This can be written as equation (2.36).

$$\begin{vmatrix} \varepsilon_x - \varepsilon_p & \dfrac{\gamma_{xy}}{2} & \dfrac{\gamma_{xz}}{2} \\[3mm] \dfrac{\gamma_{xy}}{2} & \varepsilon_y - \varepsilon_p & \dfrac{\gamma_{yz}}{2} \\[3mm] \dfrac{\gamma_{xz}}{2} & \dfrac{\gamma_{yz}}{2} & \varepsilon_z - \varepsilon_p \end{vmatrix} = 0 \tag{2.36}$$

Expanding equation (2.36) leads to the strain cubic equation (2.37). The three roots of the strain cubic equation are the principal strains; ε_1, ε_2, and ε_3. Since the principal strains are independent of the original orientation of the given strain tensor; J_1, J_2, and J_3 must also be independent of the original orientation. Thus; J_1, J_2, and J_3 are known as the strain invariants and are useful for validating strain tensor transformations. It should also be observed that the principal strains are the eigenvalues of equation (2.35) and the principal strain directions are the eigenvectors of equation (2.35).

$$\varepsilon_p^3 - J_1 \varepsilon_p^2 + J_2 \varepsilon_p - J_3 = 0 \tag{2.37}$$

Where;

$$J_1 = \varepsilon_x + \varepsilon_y + \varepsilon_z$$

$$J_2 = \varepsilon_x \varepsilon_y + \varepsilon_x \varepsilon_z + \varepsilon_y \varepsilon_z - \frac{1}{4}(\gamma_{xy}^2 + \gamma_{yz}^2 + \gamma_{xz}^2)$$

$$J_3 = \begin{vmatrix} \varepsilon_x & \dfrac{\gamma_{xy}}{2} & \dfrac{\gamma_{xz}}{2} \\[3mm] \dfrac{\gamma_{xy}}{2} & \varepsilon_y & \dfrac{\gamma_{yz}}{2} \\[3mm] \dfrac{\gamma_{xz}}{2} & \dfrac{\gamma_{yz}}{2} & \varepsilon_z \end{vmatrix}$$

In general the solution to the strain cubic equation is difficult to obtain. However, a practical approach to determining the roots of the strain cubic equation is given in Reference 4, Appendix B.

The strain cubic equation can be simplified in a 2D plane strain environment to obtain the 2D plane strain principal strains, as given by equation (2.38).

$$\varepsilon_p^2 - J_1\varepsilon_p + J_2 = 0 \qquad (2.38)$$

Where;

$$J_1 = \varepsilon_x + \varepsilon_y$$

$$J_2 = \varepsilon_x\varepsilon_y - \frac{1}{4}\gamma_{xy}^2$$

Solving equation (2.38) using the quadratic formula yields the 2D plane strain principal strains, as given in equation (2.39).

$$\varepsilon_{1,2} = \frac{\varepsilon_x + \varepsilon_y}{2} \pm \sqrt{\left(\frac{\varepsilon_x - \varepsilon_y}{2}\right)^2 + \left(\frac{\gamma_{xy}}{2}\right)^2} \qquad (2.39)$$

The octahedral shear strain is the maximum shear strain that exists in a 3D state of strain. The octahedral shear strain acts on octahedral planes relative to the principal planes. However, unlike the principal strain planes that have zero shear strain, the octahedral shear strain planes have a normal strain component, called the octahedral normal strain or the mean strain. The octahedral shear strain and the corresponding mean strain, in terms of the principal strains, are given by equations (2.40) and (2.41) respectively. The octahedral shear strain can also be written in terms of the strain tensor components, as given by equation (2.42)

$$\gamma_{oct} = \frac{2}{3}\sqrt{(\varepsilon_1 - \varepsilon_2)^2 + (\varepsilon_2 - \varepsilon_3)^2 + (\varepsilon_3 - \varepsilon_1)^2} \qquad (2.40)$$

$$\varepsilon_{mean} = \frac{1}{3}(\varepsilon_1 + \varepsilon_2 + \varepsilon_3) \qquad (2.41)$$

$$\gamma_{oct} = \frac{2}{3}\sqrt{(\varepsilon_x - \varepsilon_y)^2 + (\varepsilon_y - \varepsilon_z)^2 + (\varepsilon_z - \varepsilon_x)^2 + \frac{3}{2}(\gamma_{xy}^2 + \gamma_{yz}^2 + \gamma_{xz}^2)} \qquad (2.42)$$

The octahedral strains can be simplified in a 2D plane strain environment to obtain the 2D plane strain maximum shear strain and mean strain, as given by equations (2.43) and (2.44) respectively.

$$\gamma_{max} = (\varepsilon_1 - \varepsilon_2) \qquad (2.43)$$

$$\varepsilon_{mean} = \frac{1}{2}(\varepsilon_1 + \varepsilon_2) \qquad (2.44)$$

The von Mises strain is a simple extension of the von Mises stress in strain space. The von Mises strain in terms of the principal strains is given by equation (2.45). The von Mises strain in terms of the strain components is given by equation (2.45a).

$$\varepsilon_{vm} = \frac{1}{\sqrt{2}}\sqrt{(\varepsilon_1 - \varepsilon_2)^2 + (\varepsilon_2 - \varepsilon_3)^2 + (\varepsilon_3 - \varepsilon_1)^2} \qquad (2.45)$$

$$\varepsilon_{vm} = \frac{1}{\sqrt{2}}\sqrt{(\varepsilon_x - \varepsilon_y)^2 + (\varepsilon_y - \varepsilon_z)^2 + (\varepsilon_z - \varepsilon_x)^2 + \frac{3}{2}(\tau_{xy}^2 + \tau_{yz}^2 + \tau_{xz}^2)} \quad (2.45a)$$

The von Mises strain in 2D plane strain is given by equation (2.46).

$$\varepsilon_{vm} = \sqrt{\varepsilon_1^2 - \varepsilon_1\varepsilon_2 + \varepsilon_2^2} \qquad (2.46)$$

2.7 Generalized Hooke's Law

So far we have defined stress and strain, derived the equilibrium equations (2.22), and derived the strain-displacement equations (2.32). The equilibrium and strain-displacement equations represent nine equations for fifteen unknowns; three displacement fields u, v, w; six strain component fields ε_x, ε_y, ε_z, γ_{xy}, γ_{yz}, γ_{xz}; and six stress component fields σ_x, σ_y, σ_z, τ_{xy}, τ_{yx}, τ_{xz}. An additional six equations are therefore necessary to solve for the fifteen unknowns. It then becomes reasonable to ask if there is a relationship that can be derived between stress and strain. It turns out there is a relationship between stress and strain. For example, consider a bar in tension within the linear elastic regime. It is well know that a bar in tension within the linear elastic regime exhibits a linear relationship between the axial stress and axial strain within the bar, as given by equation (2.47).

$$\sigma_x = E\varepsilon_x \qquad (2.47)$$

The Young's Modulus, E, is a property of the material from which the bar is made. Expanding this relationship mathematically to three dimensions, we can write the relationship between stress and strain in the linear elastic regime as a linear combination of stress and strain given by equation (2.48). This equation is known as generalized Hooke's law.

$$\{\sigma\}_x = [C]\{\varepsilon\}_x$$

$$
\begin{Bmatrix} \sigma_x \\ \sigma_y \\ \sigma_z \\ \tau_{xy} \\ \tau_{yz} \\ \tau_{xz} \end{Bmatrix} =
\begin{bmatrix}
C_{11} & C_{12} & C_{13} & C_{14} & C_{15} & C_{16} \\
C_{21} & C_{22} & C_{23} & C_{24} & C_{25} & C_{26} \\
C_{31} & C_{32} & C_{33} & C_{34} & C_{35} & C_{36} \\
C_{41} & C_{42} & C_{43} & C_{44} & C_{45} & C_{46} \\
C_{51} & C_{52} & C_{53} & C_{54} & C_{55} & C_{56} \\
C_{61} & C_{62} & C_{63} & C_{64} & C_{65} & C_{66}
\end{bmatrix}
\begin{Bmatrix} \varepsilon_x \\ \varepsilon_y \\ \varepsilon_z \\ \gamma_{xy} \\ \gamma_{yz} \\ \gamma_{xz} \end{Bmatrix}
\qquad (2.48)
$$

The [C] matrix is known as the stiffness matrix of a material. The coefficients that make up the stiffness matrix, C_{ij}, are material properties of the given material in the linear elastic regime. The stiffness matrix is always a symmetric matrix for any linear elastic material. Therefore, for an anisotropic material, there are twenty-one independent C_{ij} stiffness matrix coefficients. Several material laws in the linear elastic regime are discussed in Chapter 3, in which the stiffness matrix coefficients C_{ij} are defined for each material law. The behavior of most engineering materials in the linear elastic regime can be accurately modeled using generalized Hooke's law with one of the linear elastic material laws of Chapter 3. The stiffness matrix [C] is a 4th order tensor which obeys 4th order tensor transformation laws.

2.8 General Elasticity Problem

The general elasticity problem can be solved by combining the three equilibrium equations (2.22), with the six strain-displacement equations (2.32), and the six equations that represent Hooke's law relating stress and strain (2.48) in various ways to solve for the fifteen unknown variables; three displacement fields u, v, w; six strain component fields ε_x, ε_y, ε_z, γ_{xy}, γ_{yz}, γ_{xz}; and six stress component fields σ_x, σ_y, σ_z, τ_{xy}, τ_{yx}, τ_{xz}. One such combination, assuming homogeneous isotropic material in the linear elastic regime, involves first substituting the six strain-displacement equations (2.32) with the six equations of Hooke's law (2.48) to write Hooke's law in terms of the displacement fields as given by equation (2.49).

$$
\begin{Bmatrix} \sigma_x \\ \sigma_y \\ \sigma_z \\ \tau_{xy} \\ \tau_{yz} \\ \tau_{xz} \end{Bmatrix} = \begin{bmatrix} C_{11} & C_{12} & C_{12} & 0 & 0 & 0 \\ C_{12} & C_{11} & C_{12} & 0 & 0 & 0 \\ C_{12} & C_{12} & C_{11} & 0 & 0 & 0 \\ 0 & 0 & 0 & C_{44} & 0 & 0 \\ 0 & 0 & 0 & 0 & C_{44} & 0 \\ 0 & 0 & 0 & 0 & 0 & C_{44} \end{bmatrix} \begin{Bmatrix} \dfrac{\partial u}{\partial x} \\[2mm] \dfrac{\partial v}{\partial y} \\[2mm] \dfrac{\partial w}{\partial z} \\[2mm] \dfrac{\partial u}{\partial y} + \dfrac{\partial v}{\partial x} \\[2mm] \dfrac{\partial v}{\partial z} + \dfrac{\partial w}{\partial y} \\[2mm] \dfrac{\partial u}{\partial z} + \dfrac{\partial w}{\partial x} \end{Bmatrix}
\tag{2.49}
$$

The second step involves substituting equation (2.49) into the three equilibrium equations (2.22) to arrive at the equilibrium equations written in terms of the displacement fields' u, v, and w; as given by equation (2.50).

$$
\frac{\partial}{\partial x}\left(C_{11}\frac{\partial u}{\partial x} + C_{12}\frac{\partial v}{\partial y} + C_{12}\frac{\partial w}{\partial z} \right) + \frac{\partial}{\partial y}\left(C_{44}\left(\frac{\partial u}{\partial y} + \frac{\partial v}{\partial x} \right) \right) +
$$

$$
\frac{\partial}{\partial z}\left(C_{44}\left(\frac{\partial u}{\partial z} + \frac{\partial w}{\partial x} \right) \right) + F_x = 0
$$

$$
\frac{\partial}{\partial y}\left(C_{12}\frac{\partial u}{\partial x} + C_{11}\frac{\partial v}{\partial y} + C_{12}\frac{\partial w}{\partial z} \right) + \frac{\partial}{\partial x}\left(C_{44}\left(\frac{\partial u}{\partial y} + \frac{\partial v}{\partial x} \right) \right) +
\tag{2.50}
$$

$$
\frac{\partial}{\partial z}\left(C_{44}\left(\frac{\partial v}{\partial z} + \frac{\partial w}{\partial y} \right) \right) + F_y = 0
$$

$$
\frac{\partial}{\partial z}\left(C_{12}\frac{\partial u}{\partial x} + C_{12}\frac{\partial v}{\partial y} + C_{11}\frac{\partial w}{\partial z} \right) + \frac{\partial}{\partial x}\left(C_{44}\left(\frac{\partial u}{\partial z} + \frac{\partial w}{\partial x} \right) \right) +
$$

$$
\frac{\partial}{\partial y}\left(C_{44}\left(\frac{\partial v}{\partial z} + \frac{\partial w}{\partial y} \right) \right) + F_z = 0
$$

Using the isotropic material relationships defined below, we can write the equilibrium equations in terms of the displacement fields' u, v, and w in a simplified form as given by equation (2.51). Equation (2.51) is also called Navier's equation.

$$C_{11} = 2C_{44} + C_{12} \qquad \text{(Isotropic law)}$$

$$C_{44} = G \qquad \text{(Isotropic law)}$$

$$C_{12} = \lambda = \frac{vE}{(1+v)(1-2v)} \qquad \text{(Isotropic law)}$$

$$\nabla = \frac{\partial}{\partial x} + \frac{\partial}{\partial y} + \frac{\partial}{\partial z} \qquad \text{(Mathematical definition)}$$

$$\varepsilon = \frac{\partial u}{\partial x} + \frac{\partial v}{\partial y} + \frac{\partial w}{\partial z} \qquad \text{(Mathematical definition)}$$

The goal of elasticity is then to solve the system of equations (2.51) for the three displacement fields' $u(x,y,z)$, $v(x,y,z)$, and $w(x,y,z)$, that also satisfy the boundary conditions. Once the displacement fields are determined, one can systematically use the displacement fields to calculate the strains (2.32) and the stresses (2.48) at any point (x,y,z) within a homogeneous isotropic body in the linear elastic regime. In general, it is not straightforward to solve the system of equations (2.51) for a given boundary condition; therefore, the need for the strength of materials approaches to derive the classical lamination theory equations for the analysis of laminated plates.

$$G\nabla^2 u + (G+\lambda)\frac{\partial}{\partial x}\varepsilon + Fx = 0$$

$$G\nabla^2 v + (G+\lambda)\frac{\partial}{\partial y}\varepsilon + Fy = 0 \qquad (2.51)$$

$$G\nabla^2 w + (G+\lambda)\frac{\partial}{\partial z}\varepsilon + Fz = 0$$

2.9 Vector Transformation Law

Vectors in R^3 space, all real numbers in the three-dimensional vector space, are represented in ijk notation given by equation (2.52). The vectors i, j, and k are unit vectors that are defined by equation (2.53).

$$\vec{v} = v_x i + v_y j + v_z k \tag{2.52}$$

Where;
$$i = \{1 \quad 0 \quad 0\}$$
$$j = \{0 \quad 1 \quad 0\} \tag{2.53}$$
$$k = \{0 \quad 0 \quad 1\}$$

Unit vectors are vectors with a magnitude of unity. Any general vector can be made a unit vector by utilizing equation (2.54).

$$\hat{v} = \frac{\vec{v}}{\|\vec{v}\|} \tag{2.54}$$

Where;
$$\|\vec{v}\| = \sqrt{v_x^2 + v_y^2 + v_z^2}$$

Vectors can be transformed from a given (xyz) general coordinate system to any other (x'y'z') general coordinate system according to the transformation law given in equation (2.55). The (xyz) general coordinate system is defined by three unit vectors, each one defining one of its x-, y-, and z-axis directions. The x-, y-, and z-axis direction vectors must be unit vectors. In addition, the (x'y'z') general coordinate system is also defined by three unit vectors, each one defining one of its x'-, y'-, and z'-axis directions. The x'-, y'-, and z'-axis direction vectors must also be unit vectors. The l, m, and n values in equation (2.55) and (2.56) are the direction cosines between the axis directions of the (xyz) and the (x'y'z') general coordinate systems. Specifically, the l_1, l_2, and l_3 values are the direction cosines between the x-axis and the x'-, y'-, and z'-axis directions respectively, as defined in figure 2.7. The m values are the direction cosines between the y-axis and the x'-, y'-, and z'-axis directions. Similarly, the n values are the direction cosines between the z-axis and the x'-, y'-, and z'-axis directions.

$$\begin{Bmatrix} v_{x'} \\ v_{y'} \\ v_{z'} \end{Bmatrix} = \begin{bmatrix} l_1 & m_1 & n_1 \\ l_2 & m_2 & n_2 \\ l_3 & m_3 & n_3 \end{bmatrix} \begin{Bmatrix} v_x \\ v_y \\ v_z \end{Bmatrix} \qquad (2.55)$$

$$\begin{Bmatrix} v_x \\ v_y \\ v_z \end{Bmatrix} = \begin{bmatrix} l_1 & l_2 & l_3 \\ m_1 & m_2 & m_3 \\ n_1 & n_2 & n_3 \end{bmatrix} \begin{Bmatrix} v_{x'} \\ v_{y'} \\ v_{z'} \end{Bmatrix} \qquad (2.56)$$

Where;

$l_1 = \cos(x, x') = \hat{x} \bullet \hat{x}' = \hat{x}'_x$

$l_2 = \cos(x, y') = \hat{x} \bullet \hat{y}' = \hat{y}'_x$

$l_3 = \cos(x, z') = \hat{x} \bullet \hat{z}' = \hat{z}'_x$

$m_1 = \cos(y, x') = \hat{y} \bullet \hat{x}' = \hat{x}'_y$

$m_2 = \cos(y, y') = \hat{y} \bullet \hat{y}' = \hat{y}'_y$

$m_3 = \cos(y, z') = \hat{y} \bullet \hat{z}' = \hat{z}'_y$

$n_1 = \cos(z, x') = \hat{z} \bullet \hat{x}' = \hat{x}'_z$

$n_2 = \cos(z, y') = \hat{z} \bullet \hat{y}' = \hat{y}'_z$

$n_3 = \cos(z, z') = \hat{z} \bullet \hat{z}' = \hat{z}'_z$

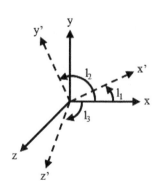

Figure 2.7, Definition of Direction Cosines for Transformation Laws

The following properties of the direction cosines (l, m, n) must be satisfied due to trigonometric relationships and the orthogonal (xyz) and (x'y'z') axis directions.

$$l_1^2 + m_1^2 + n_1^2 = 1$$
$$l_2^2 + m_2^2 + n_2^2 = 1 \qquad (2.57)$$
$$l_3^2 + m_3^2 + n_3^2 = 1$$

$$l_1 l_2 + m_1 m_2 + n_1 n_2 = 0$$
$$l_2 l_3 + m_2 m_3 + n_2 n_3 = 0 \qquad (2.58)$$
$$l_1 l_3 + m_1 m_3 + n_1 n_3 = 0$$

2.10 2nd Order Tensor Transformation Law

Stress and strain tensors are tensors of 2nd order and transform obeying the 2nd order tensor transformation law given by equation (2.59). It is important to note that the shear strains in equation (2.60) are engineering shear strains; therefore, the strain transformation law must be modified to account for the difference between engineering and tensor shear strains. If tensor shear strains are used, then equation (2.59) is valid for stress and strain tensor transformations. If engineering shear strains are used, then only equation (2.60) is valid for strain transformations. Engineering shear strain is double the tensor shear strains by definition. The l, m, and n values in equations (2.59) and (2.60) are defined exactly the same as they are in vector transformation of section 2.7.

$$\{\sigma\}_{x'} = [Ts]\{\sigma\}_x$$

$$
\begin{Bmatrix} \sigma_{x'} \\ \sigma_{y'} \\ \sigma_{z'} \\ \tau_{xy'} \\ \tau_{yz'} \\ \tau_{xz'} \end{Bmatrix}
=
\begin{bmatrix}
l_1^2 & m_1^2 & n_1^2 & 2l_1 m_1 & 2m_1 n_1 & 2l_1 n_1 \\
l_2^2 & m_2^2 & n_2^2 & 2l_2 m_2 & 2m_2 n_2 & 2l_2 n_2 \\
l_3^2 & m_3^2 & n_3^2 & 2l_3 m_3 & 2m_3 n_3 & 2l_3 n_3 \\
l_1 l_2 & m_1 m_2 & n_1 n_2 & l_1 m_2 + m_1 l_2 & m_1 n_2 + n_1 m_2 & n_1 l_2 + l_1 n_2 \\
l_2 l_3 & m_2 m_3 & n_2 n_3 & m_2 l_3 + l_2 m_3 & n_2 m_3 + m_2 n_3 & l_2 n_3 + n_2 l_3 \\
l_1 l_3 & m_1 m_3 & n_1 n_3 & l_1 m_3 + m_1 l_3 & m_1 n_3 + n_1 m_3 & n_1 l_3 + l_1 n_3
\end{bmatrix}
\begin{Bmatrix} \sigma_x \\ \sigma_y \\ \sigma_z \\ \tau_{xy} \\ \tau_{yz} \\ \tau_{xz} \end{Bmatrix}
$$

$$(2.59)$$

$$\{\varepsilon\}_{x'} = [Te]\{\varepsilon\}_x$$

$$
\begin{Bmatrix} \varepsilon_{x'} \\ \varepsilon_{y'} \\ \varepsilon_{z'} \\ \gamma_{x'y'} \\ \gamma_{y'z'} \\ \gamma_{x'z'} \end{Bmatrix} =
\begin{bmatrix}
l_1^2 & m_1^2 & n_1^2 & l_1 m_1 & m_1 n_1 & l_1 n_1 \\
l_2^2 & m_2^2 & n_2^2 & l_2 m_2 & m_2 n_2 & l_2 n_2 \\
l_3^2 & m_3^2 & n_3^2 & l_3 m_3 & m_3 n_3 & l_3 n_3 \\
2l_1 l_2 & 2m_1 m_2 & 2n_1 n_2 & l_1 m_2 + m_1 l_2 & m_1 n_2 + n_1 m_2 & n_1 l_2 + l_1 n_2 \\
2l_2 l_3 & 2m_2 m_3 & 2n_2 n_3 & m_2 l_3 + l_2 m_3 & n_2 m_3 + m_2 n_3 & l_2 n_3 + n_2 l_3 \\
2l_1 l_3 & 2m_1 m_3 & 2n_1 n_3 & l_1 m_3 + m_1 l_3 & m_1 n_3 + n_1 m_3 & n_1 l_3 + l_1 n_3
\end{bmatrix}
\begin{Bmatrix} \varepsilon_x \\ \varepsilon_y \\ \varepsilon_z \\ \gamma_{xy} \\ \gamma_{yz} \\ \gamma_{xz} \end{Bmatrix}
$$

$$(2.60)$$

2.11 4th Order Tensor Transformation Law

The stiffness matrix, [C], and the compliance matrix, [S], are both tensors of 4th order and transform obeying the 4th order tensor transformation law given by equation (2.61).

$$[\overline{C}] = [Ts][C][Te]^{-1}$$

$$(2.61)$$

It is fairly straightforward to derive the 4th order tensor transformation law utilizing generalized Hooke's law written in the (xyz) general rectangular coordinate system given by equation (2.62) and the inverse relationship of the stress and strain 2nd order transformation laws given by equations (2.63) and (2.64). Substituting equations (2.63) and (2.64) into equation (2.62), we obtain equation (2.65). Multiplying both sides of equation (2.65) by the [Ts] matrix, we obtain generalized Hooke's law written in the (x'y'z') general rectangular coordinate system, as defined by equation (2.66). Through the transformation of generalized Hooke's law from one general rectangular coordinate system to another, we have transformed the stiffness matrix [C], a 4th order tensor.

$$\{\sigma\}_x = [C]\{\varepsilon\}_x \qquad (2.62)$$

$$\{\sigma\}_x = [Ts]^{-1}\{\sigma\}_{x'} \qquad (2.63)$$

$$\{\varepsilon\}_x = [Te]^{-1}\{\varepsilon\}_{x'} \qquad (2.64)$$

$$[Ts]^{-1}\{\sigma\}_{x'} = [C][Te]^{-1}\{\varepsilon\}_{x'} \qquad (2.65)$$

$$\{\sigma\}_{x'} = [Ts][C][Te]^{-1}\{\varepsilon\}_{x'} = [\overline{C}]\{\varepsilon\}_{x'} \qquad (2.66)$$

Chapter 2
Exercises

2.1. Transform the vector, $v = \{1 \quad 1 \quad 0\}$, given in the global coordinate system into a rectangular coordinate system generated by rotating the global x-axis 45° about the global z-axis using right-hand rule.

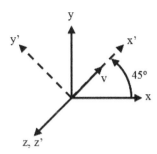

2.2. Transform the stress and strain tensors given below in the global coordinate system into a rectangular coordinate system generated by rotating the global x-axis 45° about the global z-axis using right-hand rule.

$$\{\sigma\}_x = \begin{Bmatrix} 51.5 \\ 59.0 \\ 36.5 \\ 37.6 \\ 15.0 \\ 0 \end{Bmatrix} ksi \qquad \{\varepsilon\}_x = \begin{Bmatrix} 2000 \\ 3000 \\ 0 \\ 1000 \\ 4000 \\ 0 \end{Bmatrix} \mu\varepsilon$$

2.3. Calculate the strain invariants J1, J2, J3, and ε_{vm}, for;

a) The strain tensor given in problem 2.2
b) The transformed strain tensor calculated in problem 2.2

2.4. Transform the aluminum isotropic stiffness matrix [C], given below in the global coordinate system, into a rectangular coordinate system generated by rotating the global x-axis 45° about the global z-axis using right-hand rule.

$$[C] = \begin{bmatrix} 1.48e7 & 7.30e6 & 7.30e6 & 0 & 0 & 0 \\ 7.30e6 & 1.48e7 & 7.30e6 & 0 & 0 & 0 \\ 7.30e6 & 7.30e6 & 1.48e7 & 0 & 0 & 0 \\ 0 & 0 & 0 & 3.76e6 & 0 & 0 \\ 0 & 0 & 0 & 0 & 3.76e6 & 0 \\ 0 & 0 & 0 & 0 & 0 & 3.76e6 \end{bmatrix}$$

Chapter 3
Linear Elastic Material Laws

This chapter provides a review of the equations associated with isotropic, transversely isotropic, orthotropic, and anisotropic linear elastic material laws based on generalized Hooke's law. Each of the four basic linear elastic material laws are presented along with the definition of their 3D and plane stress compliance matrix [S] and stiffness matrix [C], including restrictions on valid engineering constants for each material law.

3.1 Anisotropic Material Law

Anisotropic materials have no planes of material property symmetry; thus, the material properties are different in all directions at a specified point within a body. Anisotropic materials have twenty-one independent engineering constants, as given by equation (3.1). The anisotropic material law is better known as generalized Hooke's law.

Anisotropic Compliance Matrix

For an anisotropic material, there are thirty-six engineering constants that define the anisotropic compliance matrix given.

$$\{\varepsilon\}_{ji} = [S]\{S\}_{ji}$$

$$
\begin{Bmatrix} \varepsilon_1 \\ \varepsilon_2 \\ \varepsilon_3 \\ \gamma_{12} \\ \gamma_{23} \\ \gamma_{13} \end{Bmatrix} =
\begin{bmatrix}
S_{11} & S_{12} & S_{13} & S_{14} & S_{15} & S_{16} \\
S_{21} & S_{22} & S_{23} & S_{24} & S_{25} & S_{26} \\
S_{31} & S_{32} & S_{33} & S_{34} & S_{35} & S_{36} \\
S_{41} & S_{42} & S_{43} & S_{44} & S_{45} & S_{46} \\
S_{51} & S_{52} & S_{53} & S_{54} & S_{55} & S_{56} \\
S_{61} & S_{62} & S_{63} & S_{64} & S_{65} & S_{66}
\end{bmatrix}
\begin{Bmatrix} \sigma_1 \\ \sigma_2 \\ \sigma_3 \\ \tau_{12} \\ \tau_{23} \\ \tau_{13} \end{Bmatrix}
\qquad (3.1)
$$

However, the anisotropic compliance matrix is a symmetric matrix. Therefore, there are twenty-one independent engineering constants that define the anisotropic compliance matrix as given by equation (3.3).

$$S_{ij} = S_{ji} \tag{3.2}$$

$$\{\varepsilon\}_{ji} = [S]\{S\}_{ji}$$

$$\begin{Bmatrix} \varepsilon_1 \\ \varepsilon_2 \\ \varepsilon_3 \\ \gamma_{12} \\ \gamma_{23} \\ \gamma_{13} \end{Bmatrix} = \begin{bmatrix} S_{11} & S_{12} & S_{13} & S_{14} & S_{15} & S_{16} \\ S_{12} & S_{22} & S_{23} & S_{24} & S_{25} & S_{26} \\ S_{13} & S_{23} & S_{33} & S_{34} & S_{35} & S_{36} \\ S_{14} & S_{24} & S_{34} & S_{44} & S_{45} & S_{46} \\ S_{15} & S_{25} & S_{35} & S_{45} & S_{55} & S_{56} \\ S_{16} & S_{26} & S_{36} & S_{46} & S_{56} & S_{66} \end{bmatrix} \begin{Bmatrix} \sigma_1 \\ \sigma_2 \\ \sigma_3 \\ \tau_{12} \\ \tau_{23} \\ \tau_{13} \end{Bmatrix} \tag{3.3}$$

The twenty-one independent engineering constants that define the anisotropic compliance matrix are determined by placing a specimen of the material, figures 3.1 through 3.6, under six different stress boundary conditions, boundary conditions #1 through #6 below, within the linear elastic regime and measuring the resulting six strain tensor components under the applied stress boundary condition. The strain measurements, divided by the applied stress boundary condition, derives a column of the compliance matrix as shown in the sections below.

BC1	$\sigma_1 = \sigma_1$	$\sigma_2 = \sigma_3 = \tau_{12} = \tau_{23} = \tau_{13} = 0$	(3.4)
BC2	$\sigma_2 = \sigma_2$	$\sigma_1 = \sigma_3 = \tau_{12} = \tau_{23} = \tau_{13} = 0$	(3.8)
BC3	$\sigma_3 = \sigma_3$	$\sigma_1 = \sigma_2 = \tau_{12} = \tau_{23} = \tau_{13} = 0$	(3.12)
BC4	$\tau_{12} = \tau_{12}$	$\sigma_1 = \sigma_2 = \sigma_3 = \tau_{23} = \tau_{13} = 0$	(3.16)
BC5	$\tau_{23} = \tau_{23}$	$\sigma_1 = \sigma_2 = \sigma_3 = \tau_{12} = \tau_{13} = 0$	(3.20)
BC6	$\tau_{13} = \tau_{13}$	$\sigma_1 = \sigma_2 = \sigma_3 = \tau_{12} = \tau_{23} = 0$	(3.24)

Boundary Condition #1

Boundary condition #1, given by equation (3.4), produces the 1st column of the compliance matrix by substituting equation (3.4) into equation (3.1), producing equation (3.5).

$$\sigma_1 = \sigma_1 \tag{3.4}$$
$$\sigma_2 = \sigma_3 = \tau_{12} = \tau_{23} = \tau_{13} = 0$$

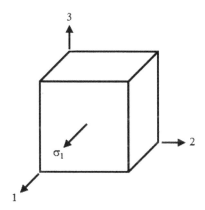

Figure 3.1, Boundary Condition #1

$$S_{11} = \frac{\varepsilon_1}{\sigma_1}, S_{21} = \frac{\varepsilon_2}{\sigma_1}, S_{31} = \frac{\varepsilon_3}{\sigma_1}, S_{41} = \frac{\gamma_{12}}{\sigma_1}, S_{51} = \frac{\gamma_{23}}{\sigma_1}, S_{61} = \frac{\gamma_{13}}{\sigma_1} \tag{3.5}$$

For a homogeneous isotropic material in the linear regime, when boundary condition #1 is applied, the following strains are produced on the cube.

$$\varepsilon_1 = \frac{\sigma_1}{E}, \varepsilon_2 = \frac{-\nu\sigma_1}{E}, \varepsilon_3 = \frac{-\nu\sigma_1}{E}, \gamma_{12} = 0, \gamma_{23} = 0, \gamma_{13} = 0 \tag{3.6}$$

Substituting equation (3.6) into (3.5) produces the 1st column of the compliance matrix for a homogeneous isotropic material in the linear elastic regime.

$$S_{11} = \frac{1}{E}, S_{21} = \frac{-\nu}{E}, S_{31} = \frac{-\nu}{E}, S_{41} = 0, S_{51} = 0, S_{61} = 0 \tag{3.7}$$

Boundary Condition #2

Boundary condition #2, given by equation (3.8), produces the 2nd column of the compliance matrix by substituting equation (3.8) into equation (3.1), producing equation (3.9).

$$\sigma_2 = \sigma_2 \qquad\qquad (3.8)$$
$$\sigma_1 = \sigma_3 = \tau_{12} = \tau_{23} = \tau_{13} = 0$$

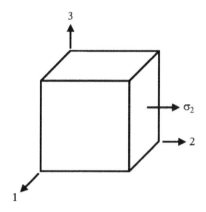

Figure 3.2, Boundary Condition #2

$$S_{12} = \frac{\varepsilon_1}{\sigma_2}, S_{22} = \frac{\varepsilon_2}{\sigma_2}, S_{32} = \frac{\varepsilon_3}{\sigma_2}, S_{42} = \frac{\gamma_{12}}{\sigma_2}, S_{52} = \frac{\gamma_{23}}{\sigma_2}, S_{62} = \frac{\gamma_{13}}{\sigma_2} \qquad (3.9)$$

For a homogeneous isotropic material in the linear regime, when boundary condition #2 is applied, the following strains are produced on the cube.

$$\varepsilon_1 = \frac{-\nu\sigma_2}{E}, \varepsilon_2 = \frac{\sigma_2}{E}, \varepsilon_3 = \frac{-\nu\sigma_2}{E}, \gamma_{12} = 0, \gamma_{23} = 0, \gamma_{13} = 0 \qquad (3.10)$$

Substituting equation (3.10) into (3.9) produces the 2nd column of the compliance matrix for a homogeneous isotropic material in the linear elastic regime.

$$S_{12} = \frac{-\nu}{E}, S_{22} = \frac{1}{E}, S_{32} = \frac{-\nu}{E}, S_{42} = 0, S_{52} = 0, S_{62} = 0 \qquad (3.11)$$

Boundary Condition #3

Boundary condition #3, given by equation (3.12), produces the 3rd column of the compliance matrix by substituting equation (3.12) into equation (3.1), producing equation (3.13).

$$\sigma_3 = \sigma_3 \tag{3.12}$$
$$\sigma_1 = \sigma_2 = \tau_{12} = \tau_{23} = \tau_{13} = 0$$

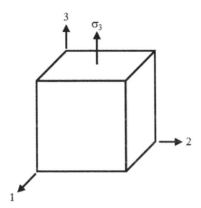

Figure 3.3, Boundary Condition #3

$$S_{13} = \frac{\varepsilon_1}{\sigma_3}, S_{23} = \frac{\varepsilon_2}{\sigma_3}, S_{33} = \frac{\varepsilon_3}{\sigma_3}, S_{43} = \frac{\gamma_{12}}{\sigma_3}, S_{53} = \frac{\gamma_{23}}{\sigma_3}, S_{63} = \frac{\gamma_{13}}{\sigma_3} \tag{3.13}$$

For a homogeneous isotropic material in the linear regime, when boundary condition #3 is applied, the following strains are produced on the cube.

$$\varepsilon_1 = \frac{-\nu\sigma_3}{E}, \varepsilon_2 = \frac{-\nu\sigma_3}{E}, \varepsilon_3 = \frac{\sigma_3}{E}, \gamma_{12} = 0, \gamma_{23} = 0, \gamma_{13} = 0 \tag{3.14}$$

Substituting equation (3.14) into (3.13) produces the 3rd column of the compliance matrix for a homogeneous isotropic material in the linear elastic regime.

$$S_{13} = \frac{-\nu}{E}, S_{23} = \frac{-\nu}{E}, S_{33} = \frac{1}{E}, S_{43} = 0, S_{53} = 0, S_{63} = 0 \tag{3.15}$$

Boundary Condition #4

Boundary condition #4, given by equation (3.16), produces the 4th column of the compliance matrix by substituting equation (3.16) into equation (3.1), producing equation (3.17).

$$\tau_{12} = \tau_{12} \tag{3.16}$$

$$\sigma_1 = \sigma_2 = \sigma_3 = \tau_{23} = \tau_{13} = 0$$

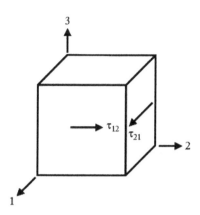

Figure 3.4, Boundary Condition #4

$$S_{14} = \frac{\varepsilon_1}{\tau_{12}}, S_{24} = \frac{\varepsilon_2}{\tau_{12}}, S_{34} = \frac{\varepsilon_3}{\tau_{12}}, S_{44} = \frac{\gamma_{12}}{\tau_{12}}, S_{45} = \frac{\gamma_{23}}{\tau_{12}}, S_{46} = \frac{\gamma_{13}}{\tau_{12}} \tag{3.17}$$

For a homogeneous isotropic material in the linear regime, when boundary condition #4 is applied, the following strains are produced on the cube.

$$\varepsilon_1 = 0, \varepsilon_2 = 0, \varepsilon_3 = 0, \gamma_{12} = \frac{\tau_{12}}{G}, \gamma_{23} = 0, \gamma_{13} = 0 \tag{3.18}$$

Substituting equation (3.18) into (3.17) produces the 4th column of the compliance matrix for a homogeneous isotropic material in the linear elastic regime.

$$S_{14} = 0, S_{24} = 0, S_{34} = 0, S_{44} = \frac{1}{G}, S_{54} = 0, S_{64} = 0 \tag{3.19}$$

Boundary Condition #5

Boundary condition #5, given by equation (3.20), produces the 5th column of the compliance matrix by substituting equation (3.20) into equation (3.1), producing equation (3.21).

$$\tau_{23} = \tau_{23} \tag{3.20}$$

$$\sigma_1 = \sigma_2 = \sigma_3 = \tau_{12} = \tau_{13} = 0$$

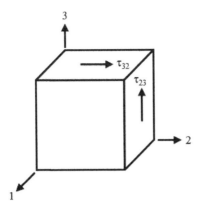

Figure 3.5, Boundary Condition #5

$$S_{15} = \frac{\varepsilon_1}{\tau_{23}}, S_{25} = \frac{\varepsilon_2}{\tau_{23}}, S_{35} = \frac{\varepsilon_3}{\tau_{23}}, S_{45} = \frac{\gamma_{12}}{\tau_{23}}, S_{55} = \frac{\gamma_{23}}{\tau_{23}}, S_{65} = \frac{\gamma_{13}}{\tau_{23}} \qquad (3.21)$$

For a homogeneous isotropic material in the linear regime, when boundary condition #5 is applied, the following strains are produced on the cube.

$$\varepsilon_1 = 0, \varepsilon_2 = 0, \varepsilon_3 = 0, \gamma_{12} = 0, \gamma_{23} = \frac{\tau_{23}}{G}, \gamma_{13} = 0 \qquad (3.22)$$

Substituting equation (3.22) into (3.21) produces the 5th column of the compliance matrix for a homogeneous isotropic material in the linear elastic regime.

$$S_{15} = 0, S_{25} = 0, S_{35} = 0, S_{45} = 0, S_{55} = \frac{1}{G}, S_{65} = 0 \qquad (3.23)$$

Boundary Condition #6

Boundary condition #6, given by equation (3.24), produces the 6th column of the compliance matrix by substituting equation (3.24) into equation (3.1), producing equation (3.25).

$$\tau_{13} = \tau_{13} \qquad (3.24)$$

$$\sigma_1 = \sigma_2 = \sigma_3 = \tau_{12} = \tau_{23} = 0$$

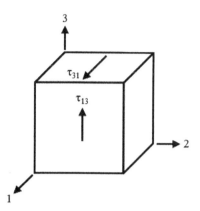

Figure 3.6, Boundary Condition #6

$$S_{16} = \frac{\varepsilon_1}{\tau_{13}}, S_{26} = \frac{\varepsilon_2}{\tau_{13}}, S_{36} = \frac{\varepsilon_3}{\tau_{13}}, S_{46} = \frac{\gamma_{12}}{\tau_{13}}, S_{56} = \frac{\gamma_{23}}{\tau_{13}}, S_{66} = \frac{\gamma_{13}}{\tau_{13}} \qquad (3.25)$$

For a homogeneous isotropic material in the linear regime, when boundary condition #6 is applied, the following strains are produced on the cube.

$$\varepsilon_1 = 0, \varepsilon_2 = 0, \varepsilon_3 = 0, \gamma_{12} = 0, \gamma_{23} = 0, \gamma_{13} = \frac{\tau_{13}}{G} \qquad (3.26)$$

Substituting equation (3.26) into (3.25) produces the 5th column of the compliance matrix for a homogeneous isotropic material in the linear elastic regime.

$$S_{16} = 0, S_{26} = 0, S_{36} = 0, S_{46} = 0, S_{56} = 0, S_{66} = \frac{1}{G} \qquad (3.27)$$

Anisotropic Stiffness Matrix

For an anisotropic material, there are thirty-six engineering constants that define the anisotropic stiffness matrix.

$$\{\sigma\}_1 = [C]\{\varepsilon\}_1$$

$$
\begin{bmatrix} \sigma_1 \\ \sigma_2 \\ \sigma_3 \\ \tau_{12} \\ \tau_{23} \\ \tau_{13} \end{bmatrix}
=
\begin{bmatrix}
C_{11} & C_{12} & C_{13} & C_{14} & C_{15} & C_{16} \\
C_{21} & C_{22} & C_{23} & C_{24} & C_{25} & C_{26} \\
C_{31} & C_{32} & C_{33} & C_{34} & C_{35} & C_{36} \\
C_{41} & C_{42} & C_{43} & C_{44} & C_{45} & C_{46} \\
C_{51} & C_{52} & C_{53} & C_{54} & C_{55} & C_{56} \\
C_{61} & C_{62} & C_{63} & C_{64} & C_{65} & C_{66}
\end{bmatrix}
\begin{bmatrix} \varepsilon_1 \\ \varepsilon_2 \\ \varepsilon_3 \\ \gamma_{12} \\ \gamma_{23} \\ \gamma_{13} \end{bmatrix}
\tag{3.28}
$$

However, the anisotropic stiffness matrix is a symmetric matrix. Therefore, there are twenty-one independent engineering constants that define the anisotropic stiffness matrix as given by equation (3.30).

$$C_{ij} = C_{ji} \tag{3.29}$$

$$\{\sigma\}_1 = [C]\{\varepsilon\}_1$$

$$
\begin{bmatrix} \sigma_1 \\ \sigma_2 \\ \sigma_3 \\ \tau_{12} \\ \tau_{23} \\ \tau_{13} \end{bmatrix}
=
\begin{bmatrix}
C_{11} & C_{12} & C_{13} & C_{14} & C_{15} & C_{16} \\
C_{12} & C_{22} & C_{23} & C_{24} & C_{25} & C_{26} \\
C_{13} & C_{23} & C_{33} & C_{34} & C_{35} & C_{36} \\
C_{14} & C_{24} & C_{34} & C_{44} & C_{45} & C_{46} \\
C_{15} & C_{25} & C_{35} & C_{45} & C_{55} & C_{56} \\
C_{16} & C_{26} & C_{36} & C_{46} & C_{56} & C_{66}
\end{bmatrix}
\begin{bmatrix} \varepsilon_1 \\ \varepsilon_2 \\ \varepsilon_3 \\ \gamma_{12} \\ \gamma_{23} \\ \gamma_{13} \end{bmatrix}
\tag{3.30}
$$

The anisotropic stiffness matrix components, C_{ij}, are determined by calculating the inverse of the anisotropic compliance matrix, $[S]$, as given by equation (3.31). Therefore, in order to determine the anisotropic stiffness matrix of a material, the anisotropic compliance matrix must first be calculated.

$$[C] = [S]^{-1} \tag{3.31}$$

Anisotropic Plane Stress Compliance Matrix

The 3-direction stresses within a body under plane stress are assumed to be zero. A state of plane stress at all points within a body is defined by equation (3.32) Thin plates, including composite laminated plates, are typically assumed to be under plane stress since their thickness is less than their width and length dimensions.

$$\sigma_3 = \tau_{23} = \tau_{13} = 0 \tag{3.32}$$

Substituting equation (3.32) into equation (3.3), the anisotropic plane stress compliance matrix, $[S^*]$, is defined.

$$\{\varepsilon\}_1 = [S^*]\{\sigma\}_1$$

$$\begin{Bmatrix} \varepsilon_1 \\ \varepsilon_2 \\ \gamma_{12} \end{Bmatrix} = \begin{bmatrix} S_{11} & S_{12} & S_{14} \\ S_{12} & S_{22} & S_{24} \\ S_{14} & S_{24} & S_{44} \end{bmatrix} \begin{Bmatrix} \sigma_1 \\ \sigma_2 \\ \tau_{12} \end{Bmatrix} \tag{3.33}$$

In addition, there exists 3-direction strains given by equation (3.34). In many element formulations, these additional 3-direction strains are typically assumed zero even though they exist in a state of plane stress as defined above.

$$\begin{Bmatrix} \varepsilon_3 \\ \gamma_{23} \\ \gamma_{13} \end{Bmatrix} = \begin{bmatrix} S_{13} & S_{23} & S_{34} \\ S_{15} & S_{25} & S_{45} \\ S_{16} & S_{16} & S_{46} \end{bmatrix} \begin{Bmatrix} \sigma_1 \\ \sigma_2 \\ \tau_{12} \end{Bmatrix} \tag{3.34}$$

Anisotropic Plane Stress Stiffness Matrix

Inverting the anisotropic plane stress compliance matrix given by equation (3.33) the anisotropic plane stress stiffness matrix, $[Q]$, is defined.

$$\{\sigma\}_1 = [S^*]^{-1}\{\varepsilon\}_1 \tag{3.35}$$

$$[Q] = [S^*]^{-1} \tag{3.36}$$

$$\{\sigma\}_1 = [Q]\{\varepsilon\}_1$$

$$\begin{Bmatrix} \sigma_1 \\ \sigma_2 \\ \tau_{12} \end{Bmatrix} = \begin{bmatrix} Q_{11} & Q_{12} & Q_{14} \\ Q_{12} & Q_{22} & Q_{24} \\ Q_{14} & Q_{24} & Q_{44} \end{bmatrix} \begin{Bmatrix} \varepsilon_1 \\ \varepsilon_2 \\ \gamma_{12} \end{Bmatrix} \tag{3.37}$$

3.2 Isotropic Material Law

Isotropic materials have an infinite number of planes of material property symmetry. Therefore, the material properties are the same in all directions at any point within a body. Isotropic materials have two independent isotropic engineering constants, any two of E, G, or v.

Isotropic Compliance Matrix

For an isotropic material, there are two independent engineering constants that define the isotropic compliance matrix.

$$\{\varepsilon\}_1 = [S]\{\sigma\}_1$$

$$\begin{Bmatrix} \varepsilon_1 \\ \varepsilon_2 \\ \varepsilon_3 \\ \gamma_{12} \\ \gamma_{23} \\ \gamma_{13} \end{Bmatrix} = \begin{bmatrix} S_{11} & S_{12} & S_{12} & 0 & 0 & 0 \\ S_{12} & S_{11} & S_{12} & 0 & 0 & 0 \\ S_{12} & S_{12} & S_{11} & 0 & 0 & 0 \\ 0 & 0 & 0 & S_{44} & 0 & 0 \\ 0 & 0 & 0 & 0 & S_{44} & 0 \\ 0 & 0 & 0 & 0 & 0 & S_{44} \end{bmatrix} \begin{Bmatrix} \sigma_1 \\ \sigma_2 \\ \sigma_3 \\ \tau_{12} \\ \tau_{23} \\ \tau_{13} \end{Bmatrix} \qquad (3.38)$$

$$S_{44} = 2(S_{11} - S_{12})$$

The isotropic compliance matrix in terms of the isotropic engineering constants is given as equation (3.39).

$$[S] = \begin{bmatrix} \dfrac{1}{E} & \dfrac{-v}{E} & \dfrac{-v}{E} & 0 & 0 & 0 \\ \dfrac{-v}{E} & \dfrac{1}{E} & \dfrac{-v}{E} & 0 & 0 & 0 \\ \dfrac{-v}{E} & \dfrac{-v}{E} & \dfrac{1}{E} & 0 & 0 & 0 \\ 0 & 0 & 0 & \dfrac{1}{G} & 0 & 0 \\ 0 & 0 & 0 & 0 & \dfrac{1}{G} & 0 \\ 0 & 0 & 0 & 0 & 0 & \dfrac{1}{G} \end{bmatrix} \qquad (3.39)$$

$$G = \frac{E}{2(1+v)} \qquad (3.40)$$

Isotropic Stiffness Matrix

For an isotropic material, there are two independent constants that define the isotropic stiffness matrix.

$$\{\sigma\}_1 = [C]\{\varepsilon\}_1$$

$$\begin{Bmatrix} \sigma_1 \\ \sigma_2 \\ \sigma_3 \\ \tau_{12} \\ \tau_{23} \\ \tau_{13} \end{Bmatrix} = \begin{bmatrix} C_{11} & C_{12} & C_{12} & 0 & 0 & 0 \\ C_{12} & C_{11} & C_{12} & 0 & 0 & 0 \\ C_{12} & C_{12} & C_{11} & 0 & 0 & 0 \\ 0 & 0 & 0 & C_{44} & 0 & 0 \\ 0 & 0 & 0 & 0 & C_{44} & 0 \\ 0 & 0 & 0 & 0 & 0 & C_{44} \end{bmatrix} \begin{Bmatrix} \varepsilon_1 \\ \varepsilon_2 \\ \varepsilon_3 \\ \gamma_{12} \\ \gamma_{23} \\ \gamma_{13} \end{Bmatrix} \qquad (3.41)$$

$$C_{44} = \frac{C_{11} - C_{12}}{2}$$

The isotropic stiffness matrix components, C_{ij}, in equation (3.20) can be determined by taking the inverse of the isotropic compliance matrix, $[S]$.

$$[C] = [S]^{-1} \qquad (3.42)$$

The isotropic stiffness matrix components in terms of the isotropic compliance matrix components, S_{ij}, are given by equation (3.43).

$$C_{11} = \frac{S_{11}^2 - S_{12}^2}{S}$$

$$C_{12} = \frac{S_{12}^2 - S_{12}S_{11}}{S} \qquad (3.43)$$

$$S = S_{11}^3 - 3S_{11}S_{12}^2 + 2S_{12}^3$$

The isotropic stiffness matrix components in terms of isotropic engineering constants are given by equation (3.44).

$$C_{11} = \frac{(1-v)E}{(1-2v)(1+v)}$$

$$C_{12} = \frac{vE}{(1-2v)(1+v)} \qquad (3.44)$$

$$C_{44} = \frac{E}{2(1+v)}$$

Isotropic Plane Stress Compliance Matrix

The state of stress at all points within a body under plane stress is defined by equation (3.45). The stresses within a body under plane stress varies only in the 1- and 2-directions, as the 3-direction stresses are assumed to be zero. Thin plates, including composite laminated plates, are typically assumed to be under a state of plane stress due to their thickness being much smaller than their width and length dimensions.

$$\sigma_3 = \tau_{23} = \tau_{23} = 0 \tag{3.45}$$

Substituting equation (3.45) into equation (3.38), the isotropic plane stress compliance matrix, $[S^*]$, is defined.

$$\{\varepsilon\}_1 = [S^*]\{\sigma\}_1$$

$$\begin{Bmatrix} \varepsilon_1 \\ \varepsilon_2 \\ \gamma_{12} \end{Bmatrix} = \begin{bmatrix} S_{11} & S_{12} & 0 \\ S_{12} & S_{11} & 0 \\ 0 & 0 & S_{44} \end{bmatrix} \begin{Bmatrix} \sigma_1 \\ \sigma_2 \\ \tau_{12} \end{Bmatrix} \tag{3.46}$$

$$S_{44} = 2(S_{11} - S_{12})$$

In addition, there exists 3-direction strains as given by equation (3.47). In many element formulations, these additional 3-direction strains are typically assumed to be zero even though they exist in a pure state of plane stress as defined by equation (3.45).

$$\varepsilon_3 = S_{12}(\sigma_1 + \sigma_2) \tag{3.47}$$

The isotropic plane stress compliance matrix in terms of the isotropic engineering constants is given by equation (3.48).

$$[S^*] = \begin{bmatrix} \dfrac{1}{E} & \dfrac{-\nu}{E} & 0 \\ \dfrac{-\nu}{E} & \dfrac{1}{E} & 0 \\ 0 & 0 & \dfrac{1}{G} \end{bmatrix} \tag{3.48}$$

Isotropic Plane Stress Stiffness Matrix

Inverting the isotropic plane stress compliance matrix given by equation (3.46), the isotropic plane stress stiffness matrix, $[Q]$, is defined.

$$\{\sigma\}_1 = [S^*]^{-1}\{\varepsilon\}_1 \tag{3.49}$$

$$[Q] = [S^*]^{-1} \tag{3.50}$$

$$\{\sigma\}_1 = [Q]\{\varepsilon\}_1$$
$$\begin{Bmatrix} \sigma_1 \\ \sigma_2 \\ \tau_{12} \end{Bmatrix} = \begin{bmatrix} Q_{11} & Q_{12} & 0 \\ Q_{12} & Q_{11} & 0 \\ 0 & 0 & Q_{44} \end{bmatrix} \begin{Bmatrix} \varepsilon_1 \\ \varepsilon_2 \\ \gamma_{12} \end{Bmatrix} \tag{3.51}$$

The isotropic plane stress stiffness matrix components, Q_{ij}, in terms of the isotropic plane stress compliance matrix components are given by equation (3.52).

$$Q_{11} = \frac{S_{11}}{S}$$
$$Q_{12} = \frac{-S_{12}}{S} \tag{3.52}$$
$$Q_{44} = \frac{1}{2(S_{11} - S_{12})}$$
$$S = S_{11}^2 - S_{12}^2$$

The isotropic plane stress stiffness matrix in terms of isotropic engineering constants is given by equation (3.53).

$$[Q] = \begin{bmatrix} \dfrac{E}{1-v^2} & \dfrac{vE}{1-v^2} & 0 \\ \dfrac{vE}{1-v^2} & \dfrac{E}{1-v^2} & 0 \\ 0 & 0 & \dfrac{E}{2(1+v)} \end{bmatrix} \tag{3.53}$$

Restrictions on Isotropic Engineering Constants

Using the isotropic engineering constant relationship given in equation (3.40) and noting that both E and G must be positive, one obtains $v > -1$. In addition, considering the stress cube of figure 2.1 under hydrostatic pressure, $\sigma_1 = \sigma_2 = \sigma_3 = -p$, and calculating the volume contraction under such a boundary condition, the hydrostatic volumetric strain, θ, is defined as follows;

$$\theta = \varepsilon_1 + \varepsilon_2 + \varepsilon_3 = \frac{-p}{\dfrac{E}{3(1-2v)}} = \frac{-p}{K} \qquad (3.54)$$

The constant K is known as the Bulk Modulus. If the Bulk Modulus is negative, then a hydrostatic pressure would cause a volumetric expansion, which is not physically possible; therefore, $v < \frac{1}{2}$. Finally, the restrictions on the isotropic engineering constants are defined.

$E > 0$

$G > 0$ $\qquad (3.55)$

$-1 < v < 1/2$

3.3 Transversely Isotropic Material Law

Transversely isotropic materials have one plane of material property symmetry in which the material properties are the same within that single plane at a specified point within a body. Transversely isotropic materials have five independent transversely isotropic engineering constants; E_1, E_2, v_{12}, v_{23}, and G_{12}. In the case of unidirectional fiber-matrix composites, the 23-plane is the single plane of material property symmetry, as shown in figure 3.7.

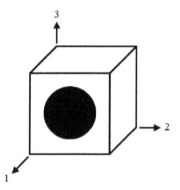

Figure 3.7, Transversely Isotropic Unidirectional Fiber-Matrix Composite

Transversely Isotropic Compliance Matrix

For a transversely isotropic material, there are five independent engineering constants that define the transversely isotropic compliance matrix.

$$\{\varepsilon\}_1 = [S]\{\sigma\}_1$$

$$
\begin{Bmatrix} \varepsilon_1 \\ \varepsilon_2 \\ \varepsilon_3 \\ \gamma_{12} \\ \gamma_{23} \\ \gamma_{13} \end{Bmatrix}
=
\begin{bmatrix}
S_{11} & S_{12} & S_{12} & 0 & 0 & 0 \\
S_{12} & S_{22} & S_{23} & 0 & 0 & 0 \\
S_{12} & S_{23} & S_{22} & 0 & 0 & 0 \\
0 & 0 & 0 & S_{44} & 0 & 0 \\
0 & 0 & 0 & 0 & 2(S_{22} - S_{23}) & 0 \\
0 & 0 & 0 & 0 & 0 & S_{44}
\end{bmatrix}
\begin{Bmatrix} \sigma_1 \\ \sigma_2 \\ \sigma_3 \\ \tau_{12} \\ \tau_{23} \\ \tau_{13} \end{Bmatrix}
\qquad (3.56)
$$

The transversely isotropic compliance matrix in terms of the transversely isotropic engineering constants is defined by equation (3.57).

$$[S] = \begin{bmatrix} \dfrac{1}{E_1} & \dfrac{-v_{21}}{E_2} & \dfrac{-v_{21}}{E_2} & 0 & 0 & 0 \\[2mm] \dfrac{-v_{12}}{E_1} & \dfrac{1}{E_2} & \dfrac{-v_{23}}{E_2} & 0 & 0 & 0 \\[2mm] \dfrac{-v_{12}}{E_1} & \dfrac{-v_{23}}{E_2} & \dfrac{1}{E_2} & 0 & 0 & 0 \\[2mm] 0 & 0 & 0 & \dfrac{1}{G_{12}} & 0 & 0 \\[2mm] 0 & 0 & 0 & 0 & \dfrac{2(1+v_{23})}{E_2} & 0 \\[2mm] 0 & 0 & 0 & 0 & 0 & \dfrac{1}{G_{12}} \end{bmatrix} \tag{3.57}$$

Due to the symmetric nature of the transversely isotropic compliance matrix, equation (3.58) can be defined. This equation is typically used to calculate v_{21} of a transversely isotropic material given v_{12}, E_1, and E_2.

$$\frac{v_{21}}{E_2} = \frac{v_{12}}{E_1} \tag{3.58}$$

Transversely Isotropic Stiffness Matrix

For a transversely isotropic material, there are five independent constants that define the transversely isotropic stiffness matrix.

$$\{\sigma\}_1 = [C]\{\varepsilon\}_1$$

$$\begin{Bmatrix} \sigma_1 \\ \sigma_2 \\ \sigma_3 \\ \tau_{12} \\ \tau_{23} \\ \tau_{13} \end{Bmatrix} = \begin{bmatrix} C_{11} & C_{12} & C_{12} & 0 & 0 & 0 \\ C_{12} & C_{22} & C_{23} & 0 & 0 & 0 \\ C_{12} & C_{23} & C_{22} & 0 & 0 & 0 \\ 0 & 0 & 0 & C_{44} & 0 & 0 \\ 0 & 0 & 0 & 0 & \dfrac{(C_{22}-C_{23})}{2} & 0 \\ 0 & 0 & 0 & 0 & 0 & C_{44} \end{bmatrix} \begin{Bmatrix} \varepsilon_1 \\ \varepsilon_2 \\ \varepsilon_3 \\ \gamma_{12} \\ \gamma_{23} \\ \gamma_{13} \end{Bmatrix} \tag{3.59}$$

The transversely isotropic stiffness matrix components, C_{ij}, can be determined by taking the inverse of the transversely isotropic compliance matrix, [S].

$$[C] = [S]^{-1} \tag{3.60}$$

The transversely isotropic stiffness matrix components in terms of the transversely isotropic compliance matrix components, S_{ij}, are given by equation (3.61).

$$C_{11} = \frac{S_{22}^2 - S_{23}^2}{S}$$

$$C_{12} = \frac{S_{12}S_{23} - S_{12}S_{22}}{S}$$

$$C_{22} = \frac{S_{11}S_{22} - S_{12}^2}{S} \tag{3.61}$$

$$C_{23} = \frac{S_{12}^2 - S_{23}S_{11}}{S}$$

$$C_{44} = \frac{1}{S_{44}}$$

$$S = S_{11}S_{22}^2 - S_{11}S_{23}^2 - 2S_{22}S_{12}^2 + 2S_{12}^2 S_{23}$$

The transversely isotropic stiffness matrix components in terms of transversely isotropic engineering constants are given by equation (3.62).

$$C = 1 - v_{23}^2 - 2(1 + v_{23})v_{12}^2 \frac{E_2}{E_1}$$

$$C_{11} = \frac{(1 - v_{23}^2)E_1}{C}$$

$$C_{12} = \frac{(1 + v_{23})v_{12}E_2}{C}$$

$$C_{22} = \frac{E_1 E_2 - v_{12}^2 E_2^2}{E_1 C} \tag{3.62}$$

$$C_{23} = \frac{v_{23}E_1 E_2 + v_{12}^2 E_2^2}{E_1 C}$$

$$C_{44} = G_{12}$$

$$\frac{C_{22} - C_{23}}{2} = \frac{E_2}{2(1 + v_{23})} = G_{23}$$

Transversely Isotropic Plane Stress Compliance Matrix

The state of stress at all points within a body under plane stress is defined by equation (3.63). The stresses within a body under plane stress varies only in the 1- and 2-directions, as the 3-direction stresses are assumed to be zero. Thin plates, including composite laminated plates, are typically assumed to be under a state of plane stress due to their thickness being much smaller than their width and length dimensions.

$$\sigma_3 = \tau_{23} = \tau_{23} = 0 \tag{3.63}$$

Substituting equation (3.63) into equation (3.56), the transversely isotropic plane stress compliance matrix, $[S^*]$, is defined.

$$\{\varepsilon\}_1 = [S^*]\{\sigma\}_1$$

$$\begin{Bmatrix} \varepsilon_1 \\ \varepsilon_2 \\ \gamma_{12} \end{Bmatrix} = \begin{bmatrix} S_{11} & S_{12} & 0 \\ S_{12} & S_{22} & 0 \\ 0 & 0 & S_{44} \end{bmatrix} \begin{Bmatrix} \sigma_1 \\ \sigma_2 \\ \tau_{12} \end{Bmatrix} \tag{3.64}$$

In addition, there exists 3-direction strains as given by equation (3.65). In many element formulations, these additional 3-direction strains are typically assumed to be zero even though they exist in a pure state of plane stress as defined by equation (3.63).

$$\varepsilon_3 = S_{12}\sigma_1 + S_{23}\sigma_2 \tag{3.65}$$

The transversely isotropic plane stress compliance matrix in terms of the transversely isotropic engineering constants is given by equation (3.66).

$$[S^*] = \begin{bmatrix} \dfrac{1}{E_1} & \dfrac{-\nu_{12}}{E_1} & 0 \\ \dfrac{-\nu_{12}}{E_1} & \dfrac{1}{E_2} & 0 \\ 0 & 0 & \dfrac{1}{G_{12}} \end{bmatrix} \tag{3.66}$$

Transversely Isotropic Plane Stress Stiffness Matrix

Inverting the transversely isotropic plane stress compliance matrix given by equation (3.64), the transversely isotropic plane stress stiffness matrix, $[Q]$, is defined.

$$\{\sigma\}_1 = [S^*]^{-1}\{\varepsilon\}_1 \tag{3.67}$$

$$[Q] = [S^*]^{-1} \tag{3.68}$$

$$\{\sigma\}_1 = [Q]\{\varepsilon\}_1$$

$$\begin{Bmatrix} \sigma_1 \\ \sigma_2 \\ \tau_{12} \end{Bmatrix} = \begin{bmatrix} Q_{11} & Q_{12} & 0 \\ Q_{12} & Q_{22} & 0 \\ 0 & 0 & Q_{44} \end{bmatrix} \begin{Bmatrix} \varepsilon_1 \\ \varepsilon_2 \\ \gamma_{12} \end{Bmatrix} \tag{3.69}$$

The transversely isotropic plane stress stiffness matrix components, Q_{ij}, in terms of the transversely isotropic plane stress compliance matrix components are given by equation (3.70).

$$Q_{11} = \frac{S_{22}}{S}$$

$$Q_{12} = \frac{-S_{12}}{S} \tag{3.70}$$

$$Q_{22} = \frac{S_{11}}{S}$$

$$Q_{44} = \frac{1}{S_{44}}$$

$$S = S_{11}S_{22} - S_{12}^2$$

The transversely isotropic plane stress stiffness matrix in terms of transversely isotropic engineering constants is given by equation (3.71).

$$[Q] = \begin{bmatrix} \dfrac{E_1}{1-\nu_{12}\nu_{21}} & \dfrac{\nu_{12}E_2}{1-\nu_{12}\nu_{21}} & 0 \\ \dfrac{\nu_{12}E_2}{1-\nu_{12}\nu_{21}} & \dfrac{E_2}{1-\nu_{12}\nu_{21}} & 0 \\ 0 & 0 & G_{12} \end{bmatrix} \tag{3.71}$$

Restrictions on Transversely Isotropic Engineering Constants

Strain energy density is defined in equation (3.72) as half the product of stress tensor times the strain tensor. Strain energy density, U_o, represents the work done by stress deforming a body per unit volume. The sum of the strain energy density over the body must be positive so as to avoid the creation of energy and violation of thermodynamic principals. In order to meet this constraint, both the transversely isotropic stiffness and compliance matrices must be positive-definite. A positive-definite transversely isotropic compliance matrix can be satisfied by applying one stress at a time, for which the corresponding strain is determined by the diagonal elements of the transversely isotropic compliance matrix. Therefore, the diagonal elements of the transversely isotropic compliance matrix must be positive so as to have a positive strain for a corresponding applied positive stress. The same holds true for the transversely isotropic stiffness matrix. These practical considerations can be expressed by writing equations (3.73) and (3.74).

$$U_o = \frac{1}{2}\sum \{\sigma\}^T \{\varepsilon\} \tag{3.72}$$

$$S_{11}, S_{22}, S_{44}, 2(S_{22}-S_{23}) > 0 \tag{3.73}$$

$$C_{11}, C_{22}. C_{44}, (C_{22}-C_{23})/2 > 0 \tag{3.74}$$

Using equations (3.73) and (3.57), the first set of restrictions on the transversely isotropic engineering constants can be derived.

$$E_1, E_2, G_{12} > 0 \tag{3.75}$$

Using equations (3.74) and (3.61), the second set of restrictions on the transversely isotropic engineering constants can be derived.

$$|v_{12}| < \sqrt{\frac{E_1}{E_2}}$$

$$-1 < v_{23} < 1 \tag{3.76}$$

$$1 - v_{23}^2 - 2(1+v_{23})v_{12}^2 \frac{E_2}{E_1} > 0$$

3.4 Orthotropic Material Law

Orthotropic materials have three orthogonal planes of material property symmetry in which the material properties are the same within these three mutually orthogonal planes at a specified point within a body. Orthotropic materials have nine independent orthotropic engineering constants, E_1, E_2, E_3, v_{12}, v_{23}, v_{13}, G_{12}, G_{23}, and G_{13}.

Orthotropic Compliance Matrix

For a transversely isotropic material, there are 9 independent engineering constants that define the orthotropic compliance matrix.

$$\{\varepsilon\}_1 = [S]\{\sigma\}_1$$

$$
\begin{Bmatrix} \varepsilon_1 \\ \varepsilon_2 \\ \varepsilon_3 \\ \gamma_{12} \\ \gamma_{23} \\ \gamma_{13} \end{Bmatrix}
=
\begin{bmatrix}
S_{11} & S_{12} & S_{13} & 0 & 0 & 0 \\
S_{12} & S_{22} & S_{23} & 0 & 0 & 0 \\
S_{13} & S_{23} & S_{33} & 0 & 0 & 0 \\
0 & 0 & 0 & S_{44} & 0 & 0 \\
0 & 0 & 0 & 0 & S_{55} & 0 \\
0 & 0 & 0 & 0 & 0 & S_{66}
\end{bmatrix}
\begin{Bmatrix} \sigma_1 \\ \sigma_2 \\ \sigma_3 \\ \tau_{12} \\ \tau_{23} \\ \tau_{13} \end{Bmatrix}
\tag{3.77}
$$

The orthotropic compliance matrix in terms of the orthotropic engineering constants is given by equation (3.78).

$$
[S] =
\begin{bmatrix}
\dfrac{1}{E_1} & \dfrac{-v_{21}}{E_2} & \dfrac{-v_{31}}{E_3} & 0 & 0 & 0 \\
\dfrac{-v_{12}}{E_1} & \dfrac{1}{E_2} & \dfrac{-v_{32}}{E_3} & 0 & 0 & 0 \\
\dfrac{-v_{13}}{E_1} & \dfrac{-v_{23}}{E_2} & \dfrac{1}{E_3} & 0 & 0 & 0 \\
0 & 0 & 0 & \dfrac{1}{G_{12}} & 0 & 0 \\
0 & 0 & 0 & 0 & \dfrac{1}{G_{23}} & 0 \\
0 & 0 & 0 & 0 & 0 & \dfrac{1}{G_{13}}
\end{bmatrix}
\tag{3.78}
$$

Due to the symmetric nature of the orthotropic compliance matrix, equation (3.79) can be defined, which is typically used to calculate v_{ji} of an orthotropic material given v_{ij}, E_i, and E_j.

$$\frac{v_{ij}}{E_i} = \frac{v_{ji}}{E_j} \tag{3.79}$$

Orthotropic Stiffness Matrix

For a transversely isotropic material, there are nine independent constants that define the orthotropic stiffness matrix.

$$\begin{Bmatrix} \sigma_1 \\ \sigma_2 \\ \sigma_3 \\ \tau_{12} \\ \tau_{23} \\ \tau_{13} \end{Bmatrix} = \begin{bmatrix} C_{11} & C_{12} & C_{13} & 0 & 0 & 0 \\ C_{12} & C_{22} & C_{23} & 0 & 0 & 0 \\ C_{13} & C_{23} & C_{33} & 0 & 0 & 0 \\ 0 & 0 & 0 & C_{44} & 0 & 0 \\ 0 & 0 & 0 & 0 & C_{55} & 0 \\ 0 & 0 & 0 & 0 & 0 & C_{66} \end{bmatrix} \begin{Bmatrix} \varepsilon_1 \\ \varepsilon_2 \\ \varepsilon_3 \\ \gamma_{12} \\ \gamma_{23} \\ \gamma_{13} \end{Bmatrix} \tag{3.80}$$

The orthotropic stiffness matrix components, C_{ij}, are determined by taking the inverse of the orthotropic compliance matrix, [S].

$$[C] = [S]^{-1} \tag{3.81}$$

The orthotropic stiffness matrix components in terms of the orthotropic compliance matrix components are given by equation (3.82).

$$C_{11} = \frac{S_{22}S_{33} - S_{23}^2}{S}$$

$$C_{12} = \frac{S_{13}S_{23} - S_{12}S_{33}}{S} \tag{3.82}$$

$$C_{13} = \frac{S_{12}S_{23} - S_{13}S_{22}}{S}$$

$$C_{22} = \frac{S_{11}S_{33} - S_{13}^2}{S}$$

$$C_{23} = \frac{S_{12}S_{13} - S_{23}S_{11}}{S}$$

$$C_{33} = \frac{S_{11}S_{22} - S_{12}^2}{S} \qquad (3.82)$$

$$C_{44} = \frac{1}{S_{44}}$$

$$C_{55} = \frac{1}{S_{55}}$$

$$C_{66} = \frac{1}{S_{66}}$$

$$S = S_{11}S_{22}S_{33} - S_{11}S_{23}^2 - S_{22}S_{13}^2 - S_{33}S_{12}^2 + 2S_{12}S_{23}S_{13}$$

Substituting the orthotropic compliance matrix components in terms of the orthotropic engineering constants from equation (3.78) into equation (3.82), the orthotropic stiffness matrix components in terms of orthotropic engineering constants are defined.

$$C = 1 - v_{12}^2 \frac{E_2}{E_1} - v_{23}^2 \frac{E_3}{E_2} - v_{13}^2 \frac{E_3}{E_1} - 2v_{12}v_{23}v_{13} \frac{E_3}{E_1}$$

$$C_{11} = \frac{(E_2 - v_{23}^2 E_3)E_1}{E_2 C}$$

$$C_{12} = \frac{v_{12}E_2 + v_{13}v_{23}E_3}{C}$$

$$C_{13} = \frac{(v_{13} + v_{12}v_{23})E_3}{C} \qquad (3.83)$$

$$C_{22} = \frac{E_1 E_2 - v_{13}^2 E_2 E_3}{E_1 C}$$

$$C_{23} = \frac{v_{23}E_1 E_3 + v_{12}v_{13}E_2 E_3}{E_1 C}$$

$$C_{33} = \frac{(E_1 - v_{12}^2 E_2)E_3}{E_1 C}$$

$$C_{44} = G_{12}$$
$$C_{55} = G_{23}$$
$$C_{66} = G_{13}$$

Orthotropic Plane Stress Compliance Matrix

The state of stress at all points within a body under plane stress is defined by equation (3.84). The stresses within a body under plane stress varies only in the 1- and 2-directions, as the 3-direction stresses are assumed to be zero. Thin plates, including composite laminated plates, are typically assumed to be under a state of plane stress due to their thickness being much smaller than their width and length dimensions.

$$\sigma_3 = \tau_{23} = \tau_{13} = 0 \tag{3.84}$$

Substituting equation (3.84) into equation (3.77), the orthotropic plane stress compliance matrix is derived.

$$\{\varepsilon\}_1 = [S^*]\{\sigma\}_1$$

$$\begin{Bmatrix} \varepsilon_1 \\ \varepsilon_2 \\ \gamma_{12} \end{Bmatrix} = \begin{bmatrix} S_{11} & S_{12} & 0 \\ S_{12} & S_{22} & 0 \\ 0 & 0 & S_{44} \end{bmatrix} \begin{Bmatrix} \sigma_1 \\ \sigma_2 \\ \tau_{12} \end{Bmatrix} \tag{3.85}$$

In addition, there exists 3-direction strains as given by equation (3.86). In many element formulations, these additional 3-direction strains are typically assumed to be zero even though they exist in a pure state of plane stress as defined by equation (3.84).

$$\varepsilon_3 = S_{13}\sigma_1 + S_{23}\sigma_2 \tag{3.86}$$

The orthotropic plane stress compliance matrix in terms of the orthotropic engineering constants is given by equation (3.87).

$$[S^*] = \begin{bmatrix} \dfrac{1}{E_1} & \dfrac{-V_{12}}{E_1} & 0 \\ \dfrac{-V_{12}}{E_1} & \dfrac{1}{E_2} & 0 \\ 0 & 0 & \dfrac{1}{G_{12}} \end{bmatrix} \tag{3.87}$$

Orthotropic Plane Stress Stiffness Matrix

Inverting the orthotropic plane stress compliance matrix given by equation (3.85) the orthotropic plane stress stiffness matrix, $[Q]$, is derived.

$$\{\sigma\}_1 = [S^*]^{-1}\{\varepsilon\}_1 \qquad (3.88)$$

$$[Q] = [S^*]^{-1} \qquad (3.89)$$

$$\{\sigma\}_1 = [Q]\{\varepsilon\}_1$$

$$\begin{Bmatrix} \sigma_1 \\ \sigma_2 \\ \tau_{12} \end{Bmatrix} = \begin{bmatrix} Q_{11} & Q_{12} & 0 \\ Q_{12} & Q_{22} & 0 \\ 0 & 0 & Q_{44} \end{bmatrix} \begin{Bmatrix} \varepsilon_1 \\ \varepsilon_2 \\ \gamma_{12} \end{Bmatrix} \qquad (3.90)$$

The orthotropic plane stress stiffness matrix components, Q_{ij}, in terms of the orthotropic plane stress compliance matrix components are given by equation (3.91).

$$Q_{11} = \frac{S_{22}}{S}$$

$$Q_{12} = \frac{-S_{12}}{S} \qquad (3.91)$$

$$Q_{22} = \frac{S_{11}}{S}$$

$$Q_{44} = \frac{1}{S_{44}}$$

$$S = S_{11}S_{22} - S_{12}^2$$

The orthotropic plane stress stiffness matrix in terms of orthotropic engineering constants is given by equation (3.92).

$$[Q] = \begin{bmatrix} \dfrac{E_1}{1-\nu_{12}\nu_{21}} & \dfrac{\nu_{12}E_2}{1-\nu_{12}\nu_{21}} & 0 \\ \dfrac{\nu_{12}E_2}{1-\nu_{12}\nu_{21}} & \dfrac{E_2}{1-\nu_{12}\nu_{21}} & 0 \\ 0 & 0 & G_{12} \end{bmatrix} \qquad (3.92)$$

Restrictions on Orthotropic Engineering Constants

Strain energy density is defined in equation (3.93) as half the product of stress tensor times the strain tensor. Strain energy density, U_o, represents the work done by stress deforming a body per unit volume. The sum of the strain energy density over the body must be positive so as to avoid the creation of energy and violation of thermodynamic principals. In order to meet this constraint, both the orthotropic stiffness and orthotropic compliance matrices must be positive-definite. A positive-definite orthotropic compliance matrix can be satisfied by applying one stress at a time, for which the corresponding strain is determined by the diagonal elements of the orthotropic compliance matrix. Therefore, the diagonal elements of the orthotropic compliance matrix must be positive so as to have a positive strain for a corresponding applied positive stress. The same holds true for the orthotropic stiffness matrix. These practical considerations can be expressed by writing equations (3.94) and (3.95).

$$U_o = \frac{1}{2}\sum \{\sigma\}^T \{\varepsilon\} > 0 \tag{3.93}$$

$$S_{11}, S_{22}. S_{33}, S_{44}, S_{55}, S_{66} > 0 \tag{3.94}$$

$$C_{11}, C_{22}. C_{33}, C_{44}, C_{55}, C_{66} > 0 \tag{3.95}$$

Using equations (3.94) and (3.78), the first set of restrictions on the orthotropic engineering constants can be determined.

$$E_1, E_2, E_3, G_{12}, G_{23}, G_{13} > 0 \tag{3.96}$$

Using equations (3.95) and (3.83), the second set of restrictions on the orthotropic engineering constants can be determined.

$$
\begin{aligned}
&|\nu_{12}| < \sqrt{\frac{E_1}{E_2}} \\[6pt]
&|\nu_{23}| < \sqrt{\frac{E_2}{E_3}} \\[6pt]
&|\nu_{13}| < \sqrt{\frac{E_1}{E_3}} \\[6pt]
&1 - \nu_{12}^2 \frac{E_2}{E_1} - \nu_{23}^2 \frac{E_3}{E_2} - \nu_{13}^2 \frac{E_3}{E_1} - 2\nu_{12}\nu_{23}\nu_{13}\frac{E_3}{E_1} > 0
\end{aligned}
\tag{3.97}
$$

3.5 Generalized Hooke's Law Including Thermal Strains

Generalized Hooke's law is given by equation (3.30). However, generalized Hooke's law needs to be modified in order to account for thermal strains. This modification is given by equation (3.98). To understand the modification to generalized Hooke's law for thermal strains, consider a bar unconstrained on a table at room temperature, as shown by figure 3.8. Now imagine that the temperature of the room is increased; the bar will expand an amount $\alpha \Delta T$ due to the temperature change. This expansion, while causing strain, will not cause the bar to have any stress as it is unconstrained on the table and allowed to freely expand. Therefore, equation (3.30) needs to be modified to account for such a condition by defining different types of strains that contribute to the overall strain of a body, as defined by equations (3.99) through (3.101). These equations define three different types of strain; the total strain $\{\varepsilon\}$, the free thermal strain $\{\varepsilon\}_t$, and the mechanical strain $\{\varepsilon\}_m$. Total strain is the summation of the free thermal strain and mechanical strain, as given by equation (3.99). The free thermal strain is the strain caused by free thermal expansion or contraction, as given by equation (3.100). The mechanical strain is the strain that causes stress, as given by equation (3.101). Typically the mechanical strain is the strain of interest in engineering calculations, as it is the strain that causes stress. However, note that most commercial FEA codes output the total strain by default; therefore, the user must be careful to set the appropriate option to output the needed mechanical strain if thermal boundary conditions are present.

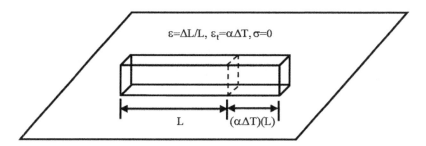

Figure 3.8, Bar under Free Thermal Expansion

$$\{\sigma\} = [C](\{\varepsilon\} - \{\alpha\}\Delta T)$$

$$
\begin{Bmatrix} \sigma_1 \\ \sigma_2 \\ \sigma_3 \\ \tau_{12} \\ \tau_{23} \\ \tau_{13} \end{Bmatrix}
=
\begin{bmatrix}
C_{11} & C_{12} & C_{13} & C_{14} & C_{15} & C_{16} \\
C_{12} & C_{22} & C_{23} & C_{24} & C_{25} & C_{26} \\
C_{13} & C_{23} & C_{33} & C_{34} & C_{35} & C_{36} \\
C_{14} & C_{24} & C_{34} & C_{44} & C_{45} & C_{46} \\
C_{15} & C_{25} & C_{35} & C_{45} & C_{55} & C_{56} \\
C_{16} & C_{26} & C_{36} & C_{46} & C_{56} & C_{66}
\end{bmatrix}
\left(
\begin{Bmatrix} \varepsilon_1 \\ \varepsilon_2 \\ \varepsilon_3 \\ \gamma_{12} \\ \gamma_{23} \\ \gamma_{13} \end{Bmatrix}
-
\begin{Bmatrix} \alpha_1 \\ \alpha_2 \\ \alpha_3 \\ \alpha_{12} \\ \alpha_{23} \\ \alpha_{13} \end{Bmatrix}
\Delta T
\right)
\qquad (3.98)
$$

Where;

$\{\varepsilon\}$ is the total strain

$\{\alpha\}$ is the coefficient of thermal expansion for the material

The total strain $\{\varepsilon\}$ is defined as:

$$\{\varepsilon\} = \{\varepsilon\}_t + \{\varepsilon\}_m \qquad (3.99)$$

The free thermal strain $\{\varepsilon\}_t$ is defined as:

$$\{\varepsilon\}_t = \{\alpha\}\Delta T \qquad (3.100)$$

The mechanical strain $\{\varepsilon\}_m$ is defined as:

$$\{\varepsilon\}_m = \{\varepsilon\} - \{\varepsilon\}_t = \{\varepsilon\} - \{\alpha\}\Delta T \qquad (3.101)$$

Using these definitions, one can write equation (3.98) in a more compact form accounting for free thermal strains. Equation (3.102) defines generalized Hooks law considering thermal strain.

$$\{\sigma\} = [C]\{\varepsilon\}_m \qquad (3.102)$$

Equation (3.102) can be inverted to write mechanical strains in terms of the compliance matrix times the stresses.

$$\{\varepsilon\}_m = [S]\{\sigma\} \qquad (3.103)$$

Chapter 3
Exercises

3.1. Given the following transversely isotropic material properties of a unidirectional ply, calculate the transverse isotropic compliance [S], stiffness [C], plane stress compliance [S*], and plane stress stiffness [Q] matrices for the ply.

$E_1 = 22.0e6$ psi
$E_2 = 1.30e6$ psi
$\nu_{12} = 0.30$
$\nu_{23} = 0.26$
$G_{12} = 0.75e6$ psi
$\alpha_1 = -0.30e-6 \ /°F$
$\alpha_2 = 18.0e-6 \ /°F$

3.2. You are given the task of modeling a composite honeycomb stiffened laminate. However, the vendor of the honeycomb material has provided E_1 stiffness only (see image below for 1-direction). Modeling the honeycomb as a homogenized transversely isotropic material where the 23-plane is the single plane of material property symmetry, and using the restrictions on the transversely isotropic engineering constants, determine the appropriate transversely isotropic engineering constants for modeling the honeycomb material.

$E_1 = 50,000$ psi $\qquad G_{23} = \dfrac{E_2}{2(1+\nu_{23})} \qquad E_2, \nu_{12}, \nu_{23}, G_{12} = \ ?$

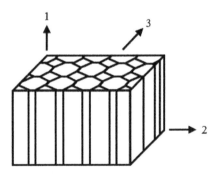

3.3. Given an isotropic aluminum bar lying on a table initially at room temperature 75°F, heat the room up to 175°F, and under the following two boundary conditions calculate the stress σ, total strain $\{\varepsilon\}$, free thermal strain $\{\varepsilon\}_t$, and mechanical strain $\{\varepsilon\}_m$ in the bar. Use isotropic aluminum properties from table 1.2.

(a) An unconstrained bar
(b) A completely constrained bar

3.4. Shown below are two plies in a laminate. Ply1 has a large coefficient of thermal expansion in the x-direction. Ply2 has a small coefficient of thermal expansion in the x-direction. For an increase in temperature, show how these plies would behave by drawing their final position. For each ply, state what the residual mechanical strains are in tension or compression.

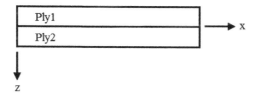

Chapter 4
Micromechanics of Composite Materials

The principal goal of micromechanics, as applied to fiber-matrix laminated composites, is to calculate the homogenized engineering constants of a ply given the engineering constants for each individual constituent that makes up the ply. In addition, micromechanics theories can be utilized to calculate the constituent mechanical stress and strain tensors given the homogenized mechanical strain tensor of the ply and the engineering constants for each constituent of the ply. This chapter reviews two micromechanics theories for the calculation of the homogenized engineering constants; the rule of mixtures theory and the modified rule of mixtures theory. In addition, this chapter reviews the phase average theory for the calculation of constituent mechanical stress and strain tensors.

4.1 Rule of Mixtures Theory

There are several micromechanics theories available that can be used to calculate the homogenized compliance matrix, [S], and the homogenized coefficients of thermal expansion, $\{\alpha\}$, of a ply. Most of the theories fall under three broad categories; empirical, strength of materials, and elasticity approaches. This section presents the rule of mixtures theory, which is the simplest of the strength of material approaches to the calculation of the homogenized engineering constants of a ply. The objective of this section is to calculate the homogenized compliance matrix, [S], and the homogenized coefficients of thermal expansion, $\{\alpha\}$, of a ply given the constituent engineering constants. In the case of fiber-matrix laminated composites, the constituents that make up a ply include fibers and matrix such that the homogenized compliance matrix and homogenized coefficients of thermal expansion are functions of the constituent engineering constants as given by equation (4.1).

$$S_{ij} = S_{ij}(E_1^f, E_2^f, v_{12}^f, v_{23}^f, G_{12}^f, V^f, E^m, v^m, V^m)$$

$$\alpha_i = \alpha_i(E_1^f, \alpha_1^f, \alpha_2^f, V^f, E^m, \alpha^m, V^m)$$

(4.1)

Fiber Properties
- E_1^f is the Young's modulus of the fiber in the 1-direction.
- E_2^f is the Young's modulus of the fiber in the 2-direction.
- v_{12}^f is the Poisson's ratio of the fiber in the 12-plane.
- v_{23}^f is the Poisson's ratio of the fiber in the 23-plane.

- $G_{12}{}^f$ is the shear modulus of the fiber in the 12-plane.
- $\alpha_1{}^f$ is the coefficient of thermal expansion of the fiber in the 1-direction.
- $\alpha_2{}^f$ is the coefficient of thermal expansion of the fiber in the 2-direction.
- V^f is the fiber volume fraction of the representative volume element.

Matrix Properties
- E^m is the Young's modulus of the matrix.
- ν^m is the Poisson's ratio of the matrix.
- α^m is the coefficient of thermal expansion of the matrix.
- V^m is the matrix volume fraction of the repeating unit cell.

The rule of mixtures theory makes the following assumptions in addition to the assumed boundary conditions applied to the representative volume element defined in each section;

- The ply is linear elastic, homogeneous, and transversely isotropic.
- The fibers are linear elastic, homogeneous, transversely isotropic or isotropic, regularly spaced, perfectly aligned and bonded to the matrix.
- The matrix is linear elastic, homogeneous, isotropic, and void free.

A representative volume element of material is the smallest region or piece of material over which the stresses and strains can be regarded as macroscopically uniform and yet the volume is still representative of the composites material and its constituents. Microscopically, however, the stresses and strains in the representative volume element are non-uniform because of the realistic heterogeneity of the composite material. Figure 4.1 shows the representative volume element for a fiber-matrix laminated composite ply used for the rule of mixtures theory. The representative volume element is used to derive the homogenized engineering constants of a ply by applying assumed stress and/or strain boundary conditions to the representative volume element in the various principal material directions. The discussion below describes the rule of mixtures approach for calculating the homogenized engineering constants of a ply; E_1, E_2, G_{12}, ν_{12}, α_1, and α_2.

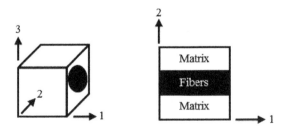

Figure 4.1, Representative Volume Element

Determining E_1 of a Transversely Isotropic Ply

Young's modulus in the 1-direction of a ply is derived by applying a strain boundary condition in the 1-direction to the representative volume, ε_1, with all other stresses zero. In addition, the assumption that the average strain in the 1-direction of the fibers and matrix is equal to the homogenized strain in the 1-direction of the ply is utilized. Therefore, the boundary conditions are defined as given in equations (4.2) and (4.3).

$$\sigma_1 = \sigma_1^f V^f + \sigma_1^m V^m, \ \sigma_2 = \tau_{12} = 0 \tag{4.2}$$

$$\varepsilon_1 = \varepsilon_1^f = \varepsilon_1^m \tag{4.3}$$

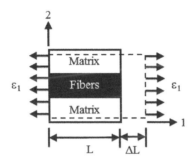

Figure 4.2, Representative Volume Element for Determination of E_1

The homogenized stress in the 1-direction of the representative volume and the average stress in the 1-direction of the fibers and matrix are defined as follows given the boundary conditions above.

$$\sigma_1 = E_1 \varepsilon_1 \tag{4.4}$$

$$\sigma_1^f = E_1^f \varepsilon_1^f \tag{4.5}$$

$$\sigma_1^m = E^m \varepsilon_1^m \tag{4.6}$$

Substituting equations (4.3) through (4.6) into equation (4.2), the rule of mixtures equation for E_1 of a transversely isotropic ply is derived.

$$E_1 = E_1^f V^f + E^m V^m \tag{4.7}$$

Determining E_2 of a Transversely Isotropic Ply

Young's modulus in the 2-direction of a ply is derived by applying a stress boundary condition in the 2-direction to the representative volume, σ_2, with all other stresses zero. In addition, the assumption that the average stress in the 2-direction of the fibers and matrix is equal to the homogenized stress in the 2-direction of the representative volume is utilized. It should be obvious that this assumption is not physically accurate. In order for the assumption to be accurate, there would have to be a strain mismatch at the matrix-fiber interface. Such a strain mismatch at the matrix-fiber interface is not consistent with strain compatibility requirements that require the strain at a material interface to be equivalent; otherwise, a crack must exist. However, determining E_2 of the homogenized ply using this assumption yields accurate enough results for studying the interaction of the constituent material properties on the homogenized ply properties. Therefore, the boundary conditions are defined as given in equations (4.8) and (4.9).

$$\varepsilon_2 = \varepsilon_2^f V^f + \varepsilon_2^m V^m \tag{4.8}$$

$$\sigma_2 = \sigma_2^f = \sigma_2^m, \quad \sigma_1 = \tau_{12} = 0 \tag{4.9}$$

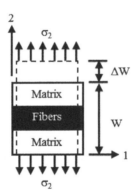

Figure 4.3, Representative Volume Element for Determination of E_2

The homogenized strain in the 2-direction of the representative volume and the average strain in the 2-direction of the fibers and matrix are defined as follows given the boundary conditions above.

$$\varepsilon_2 = \frac{\sigma_2}{E_2} \tag{4.10}$$

$$\varepsilon_2^f = \frac{\sigma_2^f}{E_2^f} = \frac{\sigma_2}{E_2^f} \tag{4.11}$$

$$\varepsilon_2^m = \frac{\sigma_2^m}{E^m} = \frac{\sigma_2}{E^m} \tag{4.12}$$

Substituting equations (4.9) through (4.12) into equation (4.8), the rule of mixtures equation for E_2 of a transversely isotropic ply is derived.

$$\frac{1}{E_2} = \frac{V^f}{E_2^f} + \frac{V^m}{E^m} \tag{4.13}$$

Or,

$$E_2 = \frac{E_2^f E^m}{E_2^f V^m + E^m V^f} \tag{4.14}$$

Determining v_{12} of a Transversely Isotropic Ply

Poisson's ratio in the 12-plane of a ply is derived by applying a strain boundary condition in the 1-direction to the representative volume, ε_1, with all other stresses zero. In addition, the assumption that the average strain in the 1-direction of the fibers and matrix is equal to the homogenized strain in the 1-direction of the representative volume is utilized. Poisson's ratio in the 12-plane is defined as the ratio of the homogenized transverse strain, ε_2, to the homogenized longitudinal strain, ε_1, as given by equation (4.15). Therefore, the boundary conditions are defined as given by equations (4.16) and (4.17).

$$v_{12} = \frac{-\varepsilon_2}{\varepsilon_1} \tag{4.15}$$

$$\varepsilon_2 = \varepsilon_2^f V^f + \varepsilon_2^m V^m \tag{4.16}$$

$$\varepsilon_1 = \varepsilon_1^f = \varepsilon_1^m \tag{4.17}$$

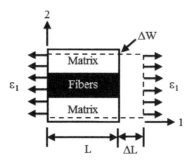

Figure 4.4, Representative Volume Element for Determination of υ_{12}

Substituting equation (4.15) into equation (4.16), the transverse strain of the representative volume can be rewritten as below.

$$v_{12}\varepsilon_1 = v_{12}^f \varepsilon_1^f V^f + v^m \varepsilon_1^m V^m \tag{4.18}$$

Substituting equation (4.17) into equation (4.18), the rule of mixtures equation for v_{12} of a transversely isotropic ply is derived.

$$v_{12} = v_{12}^f V^f + v^m V^m \tag{4.19}$$

Determination of v_{23} of a Transversely Isotropic Ply

Poisson's ratio in the 23-plane of a ply is derived by applying a stress boundary condition in the 2-direction to the representative volume, σ_2, with all other stresses zero. In addition, the assumption that average stress in the 2-direction of the fibers and matrix is equal to the homogenized stress in the 2-direction of the representative volume is utilized. It should be obvious that this assumption is not physically accurate. In order for the assumption to be accurate, there would have to be a strain mismatch at the matrix-fiber interface. Such a strain mismatch at the matrix-fiber interface is not consistent with strain compatibility requirements that require the strain at a material interface to be equivalent; otherwise, a crack must exist. However, determining v_{23} of the homogenized ply using the assumptions mentioned yields accurate enough results for studying the interaction of the constituent material properties on the effective homogenized ply properties. Poisson's ratio in the 23-plane is defined as the ratio of the homogenized through-thickness strain, ε_3, to the homogenized transverse strain, ε_2, as given by equation (4.20) below. Therefore, the boundary conditions are defined as given in equations (4.21) and (4.22).

$$v_{23} = \frac{-\varepsilon_3}{\varepsilon_2} \tag{4.20}$$

$$\varepsilon_3 = \varepsilon_3^f V^f + \varepsilon_3^m V^m \tag{4.21}$$

$$\sigma_2 = \sigma_2^f = \sigma_2^m, \ \sigma_1 = \tau_{12} = 0 \tag{4.22}$$

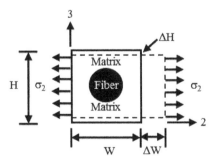

Figure 4.5, Representative Volume Element for Determination of v_{23}

Substituting equation (4.20) into equation (4.21), the through-thickness strain of the representative volume can be rewritten as follows.

$$v_{23}\varepsilon_2 = v_{23}^f\varepsilon_2^f V^f + v^m\varepsilon_2^m V^m \tag{4.23}$$

The homogenized strain in the 2-direction of the representative volume and the average strain in the 2-direction of the fibers and matrix are defined as follows given the boundary conditions above.

$$\varepsilon_2 = \frac{\sigma_2}{E_2} \tag{4.24}$$

$$\varepsilon_2^f = \frac{\sigma_2^f}{E_2^f} = \frac{\sigma_2}{E_2^f} \tag{4.25}$$

$$\varepsilon_2^m = \frac{\sigma_2^m}{E^m} = \frac{\sigma_2}{E^m} \tag{4.26}$$

Substituting equations (4.24) through (4.26) into equation (4.23), the rule of mixtures equation for v_{23} of a transversely isotropic ply is derived.

$$v_{23} = E_2 \left(\frac{v_{23}^f V^f}{E_2^f} + \frac{v^m V^m}{E^m} \right) \tag{4.27}$$

Or,

$$v_{23} = \left(\frac{v_{23}^f E^m V^f + v^m E_2^f V^m}{E^m V^f + E_2^f V^m} \right) \tag{4.28}$$

Determining G_{12} of a Transversely Isotropic Ply

The shear modulus in the 12-plane of a ply is derived by applying a shear stress boundary condition in the 12-plane to the representative volume, τ_{12}, with all other stresses zero. In addition, the assumption that the average shear stress in the 12-plane of the fibers and matrix is equal to the homogenized shear stress in the 12-plane of the representative volume is utilized. Clearly this assumption is not accurate, as the shear stress in the fiber and matrix of the representative volume cannot be the same if the shear modulus for each constituent is different. However, determination of the homogenized G_{12} of the ply using the assumption yields accurate enough results for studying the interaction of the constituent material properties on the effective homogenized ply properties. Therefore, the boundary conditions are defined as given by equations (4.29) and (4.30).

$$\gamma_{12} = \gamma_{12}^f V^f + \gamma_{12}^m V^m \tag{4.29}$$

$$\tau_{12} = \tau_{12}^f = \tau_{12}^m , \quad \sigma_1 = \sigma_2 = 0 \tag{4.30}$$

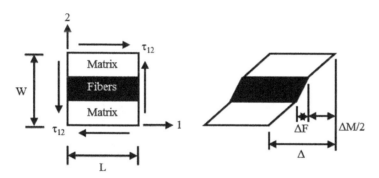

Figure 4.6, Representative Volume Element for Determination of G_{12}

The homogenized shear strain in the 12-plane of the representative volume and the average shear strain in the 12-plane of the fibers and matrix are defined as follows given the boundary conditions above.

$$\gamma_{12} = \frac{\tau_{12}}{G_{12}} \tag{4.31}$$

$$\gamma_{12}^f = \frac{\tau_{12}^f}{G_{12}^f} = \frac{\tau_{12}}{G_{12}^f} \tag{4.32}$$

$$\gamma_{12}^m = \frac{\tau_{12}^m}{G^m} = \frac{\tau_{12}}{G^m} \tag{4.33}$$

Substituting equations (4.30) through (4.33) into equation (4.29), the rule of mixtures equation for G_{12} of a transversely isotropic ply is derived.

$$\frac{1}{G_{12}} = \frac{V^f}{G_{12}^f} + \frac{V^m}{G^m}$$
(4.34)

Or,

$$G_{12} = \frac{G_{12}^f G^m}{G_{12}^f V^m + G^m V^f}$$
(4.35)

Determining α_1 of a Transversely Isotropic Ply

The coefficient of thermal expansion of a ply in the 1-direction, α_1, is determined by applying a thermal boundary condition to the representative volume. The thermal boundary condition is a constant increase in temperature of the representative volume which causes the volume to expand in all directions. In addition, the assumption that the average strain in the 1-direction of the fibers and matrix are equal to the homogenized strain in the 1-direction of the representative volume is utilized. In the case of a thermal boundary condition, the homogenized strain in the 1-direction of the representative volume is equal to $\alpha_1\Delta T$. Finally, the homogenized stress in the 1-direction of the representative volume must equal zero, as an unconstrained volume has no stress under a pure thermal load. These boundary conditions are defined as given below.

$$\sigma_1 = \sigma_1^f V^f + \sigma_1^m V^m = 0 \tag{4.36}$$

$$\varepsilon_1 = \varepsilon_1^f = \varepsilon_1^m = \alpha_1\Delta T \tag{4.37}$$

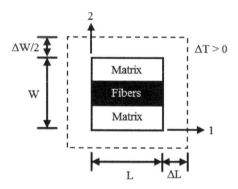

Figure 4.7, Representative Volume Element for Determination of α_1

The average stress in the 1-direction of the fibers and matrix are defined as;

$$\sigma_1^f = E_1^f(\varepsilon_1^f - \alpha_1^f \Delta T) \tag{4.38}$$

$$\sigma_1^m = E^m(\varepsilon_1^m - \alpha^m \Delta T) \tag{4.39}$$

Substituting equations (4.37) through (4.39) into equation (4.36), the rule of mixtures equation for α_1 of a transversely isotropic ply is derived.

$$\alpha_1 = \frac{\alpha_1^f E_1^f V^f + \alpha^m E^m V^m}{E_1^f V^f + E^m V^m} \tag{4.40}$$

Determining α_2 of a Transversely Isotopic Ply

The coefficient of thermal expansion in the 2-direction of a ply, α_2, is determined by applying a thermal boundary condition to the representative volume, as shown in figure 4.7. The thermal boundary condition is a constant increase in temperature applied to the representative volume which causes the volume to expand in all directions. In addition, the assumption that the average stress in the 2-direction of the fibers and matrix are equal to the homogenized stress in the 2-direction of the representative volume is utilized. The homogenized stress in the 2-direction of the representative volume must equal zero, as an unconstrained volume has no stress under a pure thermal load. Finally, the homogenized strain in the 2-direction of the representative volume is equal to $\alpha_2 \Delta T$. These boundary conditions are defined as given below.

$$\varepsilon_1 = \varepsilon_1^f = \varepsilon_1^m = \alpha_1 \Delta T \tag{4.41}$$

$$\varepsilon_2 = \varepsilon_2^f V^f + \varepsilon_2^m V^m = \alpha_2 \Delta T \tag{4.42}$$

$$\sigma_2 = \sigma_2^f = \sigma_2^m = 0 \tag{4.43}$$

The average stress in the 2-direction of the fibers and matrix are defined as;

$$\sigma_1^f = E_1^f (\varepsilon_1^f - \alpha_1^f \Delta T) \tag{4.44}$$

$$\sigma_1^m = E^m (\varepsilon_1^m - \alpha^m \Delta T) \tag{4.45}$$

The average strain in the 2-direction of the fibers and matrix are defined as;

$$\varepsilon_2^f = \frac{-v_{12}^f}{E_1^f} \sigma_1^f + \alpha_2^f \Delta T \tag{4.46}$$

$$\varepsilon_2^m = \frac{-v^m}{E^m} \sigma_1^m + \alpha^m \Delta T \tag{4.47}$$

Substituting equations (4.44) and (4.45) into equations (4.46) and (4.47);

$$\varepsilon_2^f = v_{12}^f \alpha_1^f \Delta T - v_{12}^f \alpha_1 \Delta T + \alpha_2^f \Delta T \tag{4.48}$$

$$\varepsilon_2^m = v^m \alpha^m \Delta T - v^m \alpha_1 \Delta T + \alpha^m \Delta T \tag{4.49}$$

Substituting equations (4.48) and (4.49) into equation (4.42) the rule of mixtures equation for α_2 of a transversely isotropic ply is derived.

$$\alpha_2 = \alpha_2^f V^f + \alpha^m V^m + v_{12}^f \alpha_1^f V^f + v^m \alpha^m V^m - (v_{12}^f V^f + v^m V^m)\alpha_1 \tag{4.50}$$

4.2 Modified Rule of Mixtures Theory

From the rule of mixtures theory, there were two basic assumed boundary conditions applied to the representative volume; the average strain in the 1-direction of the fiber and matrix was equivalent to the homogenized strain in the 1-direction of the representative volume, or the average stress in the 2-direction of the fibers and matrix was equivalent to the homogenized stress in the 2-direction of the representative volume. It turns out that the assumed 1-direction strain boundary condition is much more accurate to the actual boundary condition within a representative volume than is the assumed 2-direction stress boundary condition on the representative volume. Furthermore, it is observed that the predicted engineering constants that are derived from the assumed 1-direction strain boundary conditions are accurate with the actual measured engineering constants (E_1 and v_{12}). It is observed from experiment that the predicted engineering constants that are derived from the assumed 2-direction stress boundary conditions are inaccurate with the measured engineering constants (E_2, v_{23}, and G_{12}).

The largest inaccuracy of the stress boundary condition in the 2-direction is that the average stress in the 2-direction of the fibers and matrix are equal. Since the fiber is typically much stiffer than the matrix, the average stress in the 2-direction of the fibers must be larger than the average stress in the 2-direction of the matrix. To account for this fact, Tsai and Hahn conceived of a semi-empirical approach using a stress partitioning factor to relate the average fiber and matrix 2-direction stresses as given by equations (4.51) and (4.52). The stress partitioning factor, η_y, accounts for the increase in the average fiber 2-direction stress over the average matrix 2-direction stress. The shear stress partitioning factor, η_s, accounts for the increase in the average fiber 12-plane shear stress over the average matrix 12-plane shear stress. The reason the modified rule of mixtures theory is a semi-empirical approach is that the values of the stress partitioning factors should be determined from correlation with measured data. However, the shear stress partitioning factor, η_s, can actually be derived from a concentric cylinders representative volume model and is given as equation (4.54). The stress partitioning factor, η_y, does not have an analytical formula that can be derived directly; however, a value of 0.5 is commonly accepted.

$$\sigma_2^m = \eta_y \sigma_2^f, \ 0 < \eta_y \le 1 \tag{4.51}$$

$$\tau_{12}^m = \eta_s \tau_{12}^f, \ 0 < \eta_s \le 1 \tag{4.52}$$

$$\eta_y \approx 0.5 \tag{4.53}$$

$$\eta_s = \frac{1}{2}\left(1 + \frac{G^m}{G_{12}^f}\right) \tag{4.54}$$

Determining E_2 using the Stress Partitioning Factor

Young's modulus in the 2-direction of a ply is derived by applying a stress boundary condition in the 2-direction to the representative volume, σ_2, with all other stresses zero, as shown in figure 4.3. In addition, the stress partitioning factor is used to relate the average stress in the 2-direction of the fibers to the average stress in the 2-direction of the matrix. These boundary conditions are defined as given below.

$$\sigma_2 = \sigma_2^f V^f + \sigma_2^m V^m, \ \sigma_1 = \tau_{12} = 0 \tag{4.55}$$

$$\varepsilon_2 = \varepsilon_2^f V^f + \varepsilon_2^m V^m \tag{4.56}$$

Substituting equation (4.51) into equation (4.55), the homogenized stress in the 2-direction of the representative volume can be written as.

$$\sigma_2 = \sigma_2^f (V^f + \eta_y V^m) \tag{4.57}$$

Utilizing equations (4.51) and (4.57), the homogenized stress in the 2-direction of the representative volume and the average stress in the 2-direction for the fibers and matrix are defined as follows given the boundary conditions above.

$$\varepsilon_2 = \frac{\sigma_2}{E_2} = \frac{\sigma_2^f (V^f + \eta_y V^m)}{E_2} \tag{4.58}$$

$$\varepsilon_2^f = \frac{\sigma_2^f}{E_2^f} \tag{4.59}$$

$$\varepsilon_2^m = \frac{\sigma_2^m}{E^m} = \eta_y \frac{\sigma_2^f}{E^m} \tag{4.60}$$

Substituting equations (4.58) through (4.60) into equation (4.56), the modified rule of mixtures equation for E_2 of a transversely isotropic ply is derived.

$$\frac{1}{E_2} = \left(\frac{1}{V^f + \eta_y V^m} \right) \left(\frac{V^f}{E_2^f} + \eta_y \frac{V^m}{E^m} \right) \tag{4.61}$$

Or;

$$E_2 = (V^f + \eta_y V^m) \left(\frac{E_2^f E^m}{\eta_y E_2^f V^m + E^m V^f} \right) \tag{4.62}$$

Determining v_{23} using the Stress Partitioning Factor

Poisson's ratio in the 23-plane of a ply is derived by applying a stress boundary condition in the 2-direction to the representative volume with all other stresses zero. These boundary conditions are defined as given below.

$$\varepsilon_3 = \varepsilon_3^f V^f + \varepsilon_3^m V^m \tag{4.63}$$

$$\sigma_2 = \sigma_2^f V^f + \sigma_2^m V^m, \quad \sigma_1 = \tau_{12} = 0 \tag{4.64}$$

Substituting equation (4.51) into equation (4.64) the homogenized stress in the 2-direction of the representative volume can be written as;

$$\sigma_2 = \sigma_2^f (V^f + \eta_y V^m) \tag{4.65}$$

Substituting equation (4.20) into equation (4.63), the through-thickness strain of the representative volume can be written as;

$$v_{23}\varepsilon_2 = v_{23}^f \varepsilon_2^f V^f + v^m \varepsilon_2^m V^m \tag{4.66}$$

Utilizing equations (4.51) and (4.65), the homogenized strain in the 2-direction of the representative volume and the average fiber and matrix 2-direction strains are defined as follows;

$$\varepsilon_2 = \frac{\sigma_2}{E_2} = \frac{\sigma_2^f (V^f + \eta_y V^m)}{E_2} \tag{4.67}$$

$$\varepsilon_2^f = \frac{\sigma_2^f}{E_2^f} \tag{4.68}$$

$$\varepsilon_2^m = \frac{\sigma_2^m}{E^m} = \eta_y \frac{\sigma_2^f}{E^m} \tag{4.69}$$

Substituting equations (4.67) through (4.69) into equation (4.66), the modified rule of mixtures equation for v_{23} of a transversely isotropic ply is derived.

$$v_{23} = \left(\frac{E_2}{V^f + \eta_y V^m} \right) \left(\frac{v_{23}^f V^f}{E_2^f} + \eta_y \frac{v^m V^m}{E^m} \right) \tag{4.70}$$

Or,

$$v_{23} = \frac{v_{23}^f E^m V^f + \eta_y v^m E_2^f V^m}{E^m V^f + \eta_y E_2^f V^m} \tag{4.71}$$

Determining G_{12} using the Stress Partitioning Factor

The shear modulus in the 12-plane of a ply is derived by applying a shear stress boundary condition in the 12-plane to the representative volume, τ_{12}, with all other stresses zero, as shown in figure 4.6. In addition, the shear stress partitioning factor is used to relate the average shear stress in the 12-plane of the fibers to the average shear stress in the 12-plane of the matrix.

$$\tau_{12} = \tau_{12}^f V^f + \tau_{12}^m V^m \tag{4.72}$$

$$\gamma_{12} = \gamma_{12}^f V^f + \gamma_{12}^m V^m \tag{4.73}$$

Substituting equation (4.52) into equation (4.72), the homogenized shear stress in the 12-plane of the representative volume can be written as.

$$\tau_{12} = \tau_{12}^f (V^f + \eta_s V^m) \tag{4.74}$$

Utilizing equations (4.52) and (4.74) the homogenized stress in the 2-direction for the representative volume and the stress in the 2-direction for the fibers and matrix are given below as follows.

$$\gamma_{12} = \frac{\tau_{12}}{G_{12}} = \frac{\tau_{12}^f (V^f + \eta_s V^m)}{G_{12}} \tag{4.75}$$

$$\gamma_{12}^f = \frac{\tau_{12}^f}{G_{12}^f} \tag{4.76}$$

$$\gamma_{12}^m = \frac{\tau_{12}^m}{G^m} = \eta_s \frac{\tau_{12}^f}{G^m} \tag{4.77}$$

Substituting equations (4.75) through (4.77) into equation (4.73), the modified rule of mixtures equation for G_{12} of a transversely isotropic ply is derived.

$$\frac{1}{G_{12}} = \left(\frac{1}{V^f + \eta_s V^m} \right) \left(\frac{V^f}{G_{12}^f} + \eta_s \frac{V^m}{G^m} \right) \tag{4.78}$$

Or,

$$G_{12} = (V^f + \eta_s V^m) \left(\frac{G_{12}^f G^m}{\eta_s G_{12}^f V^m + G^m V^f} \right) \tag{4.79}$$

4.3 Summary of Equations

For convenience, the rule of mixtures and modified rule of mixtures equations for calculating the homogenized transversely isotropic engineering constants of a ply, given the constituent engineering constants of the fiber and matrix, are presented below. The fiber is assumed a transversely isotropic material. The matrix is assumed an isotropic material.

Summary of Rule of Mixtures Equations

$$E_1 = E_1^f V^f + E^m V^m \tag{4.80}$$

$$E_2 = \frac{E_2^f E^m}{E_2^f V^m + E^m V^f} \tag{4.81}$$

$$v_{12} = v_{12}^f V^f + v^m V^m \tag{4.82}$$

$$v_{23} = \frac{v_{23}^f E^m V^f + v^m E_2^f V^m}{E^m V^f + E_2^f V^m} \tag{4.83}$$

$$G_{12} = \frac{G_{12}^f G^m}{G_{12}^f V^m + G^m V^f} \tag{4.84}$$

Homogenized Coefficients of Thermal Expansion

$$\alpha_1 = \frac{\alpha_1^f E_1^f V^f + \alpha^m E^m V^m}{E_1^f V^f + E^m V^m} \tag{4.92}$$

$$\alpha_2 = \alpha_2^f V^f + \alpha^m V^m + v_{12}^f \alpha_1^f V^f + v^m \alpha^m V^m - (v_{12}^f V^f + v^m V^m)\alpha_1 \tag{4.93}$$

Summary of Modified Rule of Mixtures Equations

$$E_1 = E_1^f V^f + E^m V^m \tag{4.85}$$

$$E_2 = (V^f + \eta_y V^m)\left(\frac{E_2^f E^m}{\eta_y E_2^f V^m + E^m V^f}\right) \tag{4.86}$$

$$v_{12} = v_{12}^f V^f + v^m V^m \tag{4.87}$$

$$v_{23} = \frac{v_{23}^f E^m V^f + \eta_y v^m E_2^f V^m}{E^m V^f + \eta_y E_2^f V^m} \tag{4.88}$$

$$G_{12} = (V^f + \eta_s V^m)\left(\frac{G_{12}^f G^m}{\eta_s G_{12}^f V^m + G^m V^f}\right) \tag{4.89}$$

Where;

$$\eta_y \approx 0.5 \tag{4.90}$$

$$\eta_s = \frac{1}{2}\left(1 + \frac{G^m}{G_{12}^f}\right) \tag{4.91}$$

Homogenized Coefficients of Thermal Expansion

$$\alpha_1 = \frac{\alpha_1^f E_1^f V^f + \alpha^m E^m V^m}{E_1^f V^f + E^m V^m} \tag{4.92}$$

$$\alpha_2 = \alpha_2^f V^f + \alpha^m V^m + v_{12}^f \alpha_1^f V^f + v^m \alpha^m V^m - (v_{12}^f V^f + v^m V^m)\alpha_1 \tag{4.93}$$

4.4 Phase Average Theory

The intention of the phase average theory is to calculate the average stress and mechanical strain tensors for each constituent that makes up the representative volume of a ply, given the homogenized mechanical strain tensor of the ply. It is important to realize that the homogenized mechanical strain tensor of the ply, $\{\varepsilon\}_m$, represents the average mechanical strain of all the constituents that make up the representative volume of the ply. In the case of fiber-matrix laminated composites, this includes all the fibers and matrix within the representative volume. In a similar manner, the average mechanical strain tensor of a given constituent, $\{\varepsilon\}_m^f$ or $\{\varepsilon\}_m^m$, represents the average mechanical strain over some effective volume of the microstructure of the constituent.

Generalized Hooke's law for a homogenized ply material accounting for thermal strains is given by;

$$\{\sigma\} = [C]\{\varepsilon\}_m = [C](\{\varepsilon\} - \{\alpha\}\Delta T) \tag{4.94}$$

In the case of a fiber-matrix laminated composite, generalized Hooke's law for the fiber constituent accounting for thermal strains is given by;

$$\{\sigma\}^f = [C]^f\{\varepsilon\}_m^f = [C]^f(\{\varepsilon\}^f - \{\alpha\}^f\Delta T) \tag{4.95}$$

In addition, generalized Hooke's law for the matrix constituent accounting for thermal strains is given by;

$$\{\sigma\}^m = [C]^m\{\varepsilon\}_m^m = [C]^m(\{\varepsilon\}^m - \{\alpha\}^m\Delta T) \tag{4.96}$$

The phase average theory states that the homogenized stress and strain tensors of a ply are equal to the sum of the volume fraction weightings of the average stress and strain tensors of the fibers and matrix as given in equations (4.97) and (4.98).

$$\{\sigma\} = \{\sigma\}^m V^m + \{\sigma\}^f V^f \tag{4.97}$$

$$\{\varepsilon\} = \{\varepsilon\}^m V^m + \{\varepsilon\}^f V^f \tag{4.98}$$

The strain tensors in equation (4.98) are total strain tensors, since there is no subscript identification associated with the contracted strain tensor notation.

Substituting equations (4.95) and (4.96) into equation (4.97), we obtain;

$$[C]\{\varepsilon\}_m = [C]^m \{\varepsilon\}_m^m V^m + [C]^f \{\varepsilon\}_m^f V^f \qquad (4.99)$$

Expanding equation (4.98) into mechanical and free thermal strains using equation (3.55), we obtain;

$$\{\varepsilon\}_m^f V^f = \{\varepsilon\}_m - \{\varepsilon\}_m^m V^m + (\{\alpha\} - \{\alpha\}^m V^m - \{\alpha\}^f V^f) \Delta T \qquad (4.100)$$

Finally, substituting equation (4.100) into equation (4.99), the average mechanical strain tensor of the matrix is defined.

$$\{\varepsilon\}_m^m = [M]^m \{\varepsilon\}_m - \{T\}^m \Delta T \qquad (4.101)$$

Where;

$$[M]^m = \frac{1}{V^m}([C]^m - [C]^f)^{-1}([C] - [C]^f)$$

$$\{T\}^m = \frac{1}{V^m}([C]^m - [C]^f)^{-1}[C]^f(\{\alpha\} - \{\alpha\}^m V^m - \{\alpha\}^f V^f)$$

You should notice from equation (4.101) that the average mechanical strain of the matrix is a combination of a multiplying factor to the homogenized mechanical strain tensor of the ply, which includes thermal effects at the ply level due to ply orientation mismatch in coefficient of thermal expansions, $[M]^m\{\varepsilon\}_m$, plus the addition of thermal effects at the constituent level due to fiber-matrix mismatch in coefficient of thermal expansions, $\{T\}^m \Delta T$. In this way, the average mechanical strain tensor of the matrix more accurately represents the "true" strain environment of the matrix constituent and should be the preferred strain in any limit capability calculations of the ply.

In a similar manner to the average mechanical strain tensor of the matrix, the average mechanical strain tensor of the fiber is defined as follows.

$$\{\varepsilon\}_m^f = [M]^f \{\varepsilon\}_m - \{T\}^f \Delta T \qquad (4.102)$$

Where;

$$[M]^f = \frac{1}{V^f}([C]^f - [C]^m)^{-1}([C]-[C]^m)$$

$$\{T\}^f = \frac{1}{V^f}([C]^f - [C]^m)^{-1}[C]^m(\{\alpha\} - \{\alpha\}^m V^m - \{\alpha\}^f V^f)$$

Alternatively, the average mechanical strain tensor of the fiber can be determined from the average mechanical strain tensor of the matrix by utilizing equation (4.98) and expanding the total strain tensors into mechanical and free thermal strain tensors for the fiber and matrix to obtain;

$$\{\varepsilon\}_m^f = \frac{1}{V^f}(\{\varepsilon\}_m - \{\varepsilon\}_m^m V_m) + \frac{\Delta T}{V_f}(\{\alpha\} - \{\alpha\}_m V_m + \{\alpha\}_f V_f) \qquad (4.103)$$

Once the average mechanical strain tensors of the fibers and matrix have been determined, the average stress tensors of the fibers and matrix can be calculated as given in equations (4.104) and (4.105).

$$\{\sigma\}^f = [C]^f \{\varepsilon\}_m^f \qquad (4.104)$$

$$\{\sigma\}^m = [C]^m \{\varepsilon\}_m^m \qquad (4.105)$$

Chapter 4
Exercises

4.1. Calculate the homogeneous transversely isotropic engineering constants $(E_1, E_2, \nu_{12}, \nu_{23}, G_{12}, \alpha_1, \alpha_2)$ for a unidirectional ply material that is made up of the following fiber and matrix constituent material properties with a fiber volume $V^f = 0.6$; using,

(a) Rule of mixtures theory
(b) Modified rule of mixtures theory

Fiber Material Properties
$E_1^f = 36.5e6$ psi
$E_2^f = 2.4e6$ psi
$\nu_{12}^f = 0.30$
$\nu_{23}^f = 0.20$
$G_{12}^f = 6.0e6$ psi
$\alpha_1^f = -0.6e{-}6$
$\alpha_2^f = 4.0e{-}6$

Matrix Material Properties
$E^m = 0.55e6$ psi
$\nu^m = 0.30$
$\alpha^m = 30.0e{-}6$

4.2. A unidirectional ply material is to be manufactured using two fiber systems and one matrix system. How would you estimate the value of E_1 of the unidirectional ply material?

4.3. Is the homogenized transverse modulus E_2 of a unidirectional ply; stiffer, equal to, or softer than its constituent matrix modulus E_m? Provide an explanation for your answer.

4.4. Use the phase average theory to calculate;

(a) The fiber mechanical strain amplification matrix $[M]^f$ and thermal superposition vector $\{T\}^f$
(b) The matrix mechanical strain amplification matrix $[M]^m$ and thermal superposition vector $\{T\}^m$

Use the fiber and matrix constituent material properties from problem 4.1.
Use the unidirectional ply material properties from problem 3.1.

4.5. Given a unit homogeneous mechanical strain tensor with no thermal environment at some point within a unidirectional ply made from the constituent materials of problem 4.4; calculate,

(a) The average mechanical strain tensor in the fibers, $\{\varepsilon\}_m^f$
(b) The average mechanical strain tensor in the matrix, $\{\varepsilon\}_m^m$

$$\{\varepsilon\}_m = \begin{Bmatrix} 1 \\ 1 \\ 1 \\ 1 \\ 1 \\ 1 \end{Bmatrix}, \Delta T = 0, \{\varepsilon\}_m^f = ?, \{\varepsilon\}_m^m = ?$$

4.6. Given a $[0]_{10}$ laminate made from the unidirectional ply and constituent materials of problem 4.4, under a thermal environment of -100°F only; calculate,

(a) The average mechanical strain tensor in the fibers, $\{\varepsilon\}_m^f$
(b) The average mechanical strain tensor in the matrix, $\{\varepsilon\}_m^m$

$\{\varepsilon\}_m = 0, \Delta T = -100, \{\varepsilon\}_m^f = ?, \{\varepsilon\}_m^m = ?$

4.7. What do the mechanical strain amplification matrices, $[M]^f$ and $[M]^m$, account for?

4.8. What do the thermal superposition vectors, $\{T\}^f$ and $\{T\}^m$, account for?

Chapter 5
Macromechanical Behavior of a Ply

The principal objective of this chapter is to study the behavior of a single ply in the global rectangular coordinate system under plane stress conditions as the ply fiber orientation, θ, varies from $-90°$ to $90°$ as defined in figure 5.1. The stress-strain relationship, defined by generalized Hooke's law given in chapter 3, is written in the material coordinate system. This is intuitive since the material coordinate system is the coordinate system in which the engineering constants can easily be determined for a given material (i.e. the S_{ij} coefficients of the compliance matrix). However, in order to determine the behavior of a single ply with a general fiber orientation relative to the global coordinate system, we need to transform the stress-strain relationship in the material coordinate system into the global coordinate system. This transformation is accomplished using the 4th order tensor transformation law applied to the compliance and stiffness matrix of the ply appropriately.

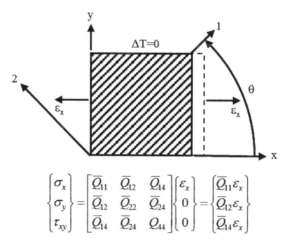

$$\begin{Bmatrix} \sigma_x \\ \sigma_y \\ \tau_{xy} \end{Bmatrix} = \begin{bmatrix} \overline{Q}_{11} & \overline{Q}_{12} & \overline{Q}_{14} \\ \overline{Q}_{12} & \overline{Q}_{22} & \overline{Q}_{24} \\ \overline{Q}_{14} & \overline{Q}_{24} & \overline{Q}_{44} \end{bmatrix} \begin{Bmatrix} \varepsilon_x \\ 0 \\ 0 \end{Bmatrix} = \begin{Bmatrix} \overline{Q}_{11}\varepsilon_x \\ \overline{Q}_{12}\varepsilon_x \\ \overline{Q}_{14}\varepsilon_x \end{Bmatrix}$$

Figure 5.1, Plane Stress in a Single Ply

5.1 Plane Stress Stiffness Matrix in Global Coordinates

In order to derive the plane stress stiffness matrix in the global coordinate system, we must first define the 2nd order tensor transformation equations under plane stress conditions. A state of plane stress is defined as $\sigma_z = \tau_{xz} = \tau_{yz} = 0$. The definition of a ply with a general fiber orientation relative to the global coordinate system under a state of plane stress is given in figure 5.1. The 2nd order tensor transformation equations (2.59) and (2.60) under a state of plane stress and transforming about the material system 3-axis direction simplify to equations (5.1) and (5.2) respectively. It is important to note that, since we are interested in transforming compliance and stiffness matrices from the material coordinate system into the global coordinate system, we choose to write the transformation law with the inverse transformation matrices to be consistent with the definitions of equations (2.59) and (2.60). Notice that the positive direction of θ defined in figure 5.1 is opposite that which is defined in the transformation equations (2.59) and (2.60); further necessitating the use of the inverse transformation matrices for consistency.

$$\{\sigma\}_x = [Ts]^{-1}\{\sigma\}_1$$

$$\begin{Bmatrix} \sigma_x \\ \sigma_y \\ \tau_{xy} \end{Bmatrix} = \begin{bmatrix} \cos^2\theta & \sin^2\theta & -2\cos\theta\sin\theta \\ \sin^2\theta & \cos^2\theta & 2\cos\theta\sin\theta \\ \cos\theta\sin\theta & -\cos\theta\sin\theta & \cos^2\theta - \sin^2\theta \end{bmatrix} \begin{Bmatrix} \sigma_1 \\ \sigma_2 \\ \tau_{12} \end{Bmatrix} \quad (5.1)$$

$$\{\varepsilon\}_x = [Te]^{-1}\{\varepsilon\}_1 = [Ts]^T\{\varepsilon\}_1$$

$$\begin{Bmatrix} \varepsilon_x \\ \varepsilon_y \\ \gamma_{xy} \end{Bmatrix} = \begin{bmatrix} \cos^2\theta & \sin^2\theta & -\cos\theta\sin\theta \\ \sin^2\theta & \cos^2\theta & \cos\theta\sin\theta \\ 2\cos\theta\sin\theta & -2\cos\theta\sin\theta & \cos^2\theta - \sin^2\theta \end{bmatrix} \begin{Bmatrix} \varepsilon_1 \\ \varepsilon_2 \\ \gamma_{12} \end{Bmatrix} \quad (5.2)$$

Taking the inverse of equations (5.1) and (5.2), the transformation equations from the global coordinate system to the material coordinate system are given as equations (5.3) and (5.4).

$$\{\sigma\}_1 = [Ts]\{\sigma\}_x$$

$$[Ts] = \begin{bmatrix} \cos^2\theta & \sin^2\theta & 2\cos\theta\sin\theta \\ \sin^2\theta & \cos^2\theta & -2\cos\theta\sin\theta \\ -\cos\theta\sin\theta & \cos\theta\sin\theta & \cos^2\theta - \sin^2\theta \end{bmatrix} \quad (5.3)$$

$$\{\varepsilon\}_1 = [Te]\{\varepsilon\}_x = [Ts]^{-1}\{\varepsilon\}_x$$

$$[Te] = \begin{bmatrix} \cos^2\theta & \sin^2\theta & \cos\theta\sin\theta \\ \sin^2\theta & \cos^2\theta & -\cos\theta\sin\theta \\ -2\cos\theta\sin\theta & 2\cos\theta\sin\theta & \cos^2\theta-\sin^2\theta \end{bmatrix} \tag{5.4}$$

The stress-strain relationship defined by Hooke's law for an isotropic, transversely isotropic, or orthotropic linear elastic material under plane stress boundary conditions, including a thermal boundary condition, is repeated below as equation (5.5).

$$\{\sigma\}_1 = [Q](\{\varepsilon\}_1 - \{\alpha\}_1\Delta T)$$

$$\begin{Bmatrix} \sigma_1 \\ \sigma_2 \\ \tau_{12} \end{Bmatrix} = \begin{bmatrix} Q_{11} & Q_{12} & 0 \\ Q_{12} & Q_{22} & 0 \\ 0 & 0 & Q_{44} \end{bmatrix} \left(\begin{Bmatrix} \varepsilon_1 \\ \varepsilon_2 \\ \gamma_{12} \end{Bmatrix} - \begin{Bmatrix} \alpha_1 \\ \alpha_2 \\ \alpha_{12} \end{Bmatrix} \Delta T \right) \tag{5.5}$$

It should be observed that equation (5.5) is written in the principal material coordinate system. However, we are interested in the behavior of a single ply in the global coordinate system. Therefore, we must transform equation (5.5) into the global coordinate system. This is achieved by substituting equations (5.1) through (5.4) into equation (5.5) to obtain;

$$\{\sigma\}_x = [Ts]^{-1}[Q][Te](\{\varepsilon\}_x - \{\alpha\}_x\Delta T) \tag{5.6}$$

It can be shown that $[Te] = [Ts]^{-T}$, which is the transpose of the matrix $[Ts]^{-1}$. Therefore, equation (5.6) can be written as equation (5.7).

$$\{\sigma\}_x = [\overline{Q}](\{\varepsilon\}_x - \{\alpha\}_x\Delta T)$$

$$\begin{Bmatrix} \sigma_x \\ \sigma_y \\ \tau_{xy} \end{Bmatrix} = \begin{bmatrix} \overline{Q}_{11} & \overline{Q}_{12} & \overline{Q}_{14} \\ \overline{Q}_{12} & \overline{Q}_{22} & \overline{Q}_{24} \\ \overline{Q}_{14} & \overline{Q}_{24} & \overline{Q}_{44} \end{bmatrix} \left(\begin{Bmatrix} \varepsilon_x \\ \varepsilon_y \\ \gamma_{xy} \end{Bmatrix} - \begin{Bmatrix} \alpha_x \\ \alpha_y \\ \alpha_{xy} \end{Bmatrix} \Delta T \right) \tag{5.7}$$

Where;
$$[\overline{Q}] = [Ts]^{-1}[Q][Ts]^{-T}$$

$[\overline{Q}]$ is the plane stress stiffness matrix transformed into the global coordinate system. The plane stress stiffness matrix in the global coordinate system defined by equation (5.7) is written out in long form below as equation (5.8).

$$\overline{Q}_{11} = Q_{11}\cos^4\theta + 2(Q_{12} + 2Q_{44})\sin^2\theta\cos^2\theta + Q_{22}\sin^4\theta$$
$$\overline{Q}_{12} = (Q_{11} + Q_{22} - 4Q_{44})\sin^2\theta\cos^2\theta + Q_{12}(\sin^4\theta + \cos^4\theta)$$
$$\overline{Q}_{14} = (Q_{11} - Q_{12} - 2Q_{44})\sin\theta\cos^3\theta + (Q_{12} - Q_{22} + 2Q_{44})\sin^3\theta\cos\theta$$
$$\overline{Q}_{22} = Q_{11}\sin^4\theta + 2(Q_{12} + 2Q_{44})\sin^2\theta\cos^2\theta + Q_{22}\cos^4\theta \qquad (5.8)$$
$$\overline{Q}_{24} = (Q_{11} - Q_{12} - 2Q_{44})\sin^3\theta\cos\theta + (Q_{12} - Q_{22} + 2Q_{44})\sin\theta\cos^3\theta$$
$$\overline{Q}_{44} = (Q_{11} + Q_{22} - 2Q_{12} - 2Q_{44})\sin^2\theta\cos^2\theta + Q_{44}(\sin^4\theta + \cos^4\theta)$$

In addition, one can invert the plane stress stiffness matrix in the global coordinate system to obtain the plane stress compliance matrix in the global coordinate system.

$$[\overline{S}] = [\overline{Q}]^{-1} \qquad (5.9)$$

$$\{\varepsilon\}_x = [\overline{S}]\{\sigma\}_x + \{\alpha\}_x\Delta T \qquad (5.10)$$

In order to study the behavior of a single ply in the global coordinate system, we need to study the components of the plane strain stiffness matrix in the global coordinate system. Figure 5.2 shows a plot of each $[\overline{Q}]$ component versus θ, the angle a ply fiber direction makes with the global x-axis as defined in figure 5.1, as θ varies from $-90°$ to $90°$. Looking at the plot of figure 5.2 the following statements can be made about the each $[\overline{Q}]$ component for a given ply.

- $\overline{Q}_{11}$ and $\overline{Q}_{22}$ are ninety degrees out of phase. Therefore, a $90°$ ply behaves as a $0°$ ply in the global y-axis direction and vice versa.
- $\overline{Q}_{11}$ and $\overline{Q}_{22}$ are always positive. Therefore, a positive extensional strain will always cause a positive extensional stress.
- $\overline{Q}_{12}$ is always positive. Therefore, a positive extensional strain in the global y-axis direction will cause a decrease in the extensional strain in the global x-axis direction due to Poisson's ratio, which requires that additional extensional stress be applied in the global x-axis direction to maintain the given extensional strain in the global x-axis direction.

- $\overline{Q}_{44}$ is always positive. Therefore, a positive shear strain will cause a positive shear stress.

- $\overline{Q}_{14}$ and $\overline{Q}_{24}$ represent extension-shear coupling terms. They are positive for positive theta and negative for negative theta. Therefore, a positive shear strain will cause additional tension stress and a negative shear strain will cause reduced tension stress for positive angle plies. The reverse is true for negative angle plies. This makes sense, as for a positive angle ply under positive shear the fibers need to elongate, whereas the same positive angle ply under negative shear will cause a shorting of the fibers. In addition, $\overline{Q}_{14}$ and $\overline{Q}_{24}$ components are zero for theta equals 0° and 90°. This means extension-shear coupling does not exist for 0° and 90° plies.

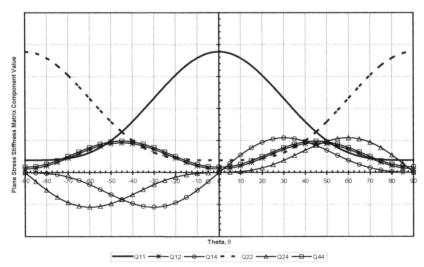

$$\left\{\begin{array}{c} \sigma_x \\ \sigma_y \\ \tau_{xy} \end{array}\right\} = \left[\begin{array}{ccc} \overline{Q}_{11} & \overline{Q}_{12} & \boxed{\overline{Q}_{14}} \\ \overline{Q}_{12} & \overline{Q}_{22} & \boxed{\overline{Q}_{24}} \\ \boxed{\overline{Q}_{14} \quad \overline{Q}_{24}} & \overline{Q}_{44} \end{array}\right]\left(\left\{\begin{array}{c} \varepsilon_x \\ \varepsilon_y \\ \gamma_{xy} \end{array}\right\} - \left\{\begin{array}{c} \alpha_x \\ \alpha_y \\ \alpha_{xy} \end{array}\right\}\Delta T\right)$$

Extension-Shear Coupling

Plane Stress Stiffness Matrix in Global Coordinates vs. Theta

Plane Stress Stiffness Matrix Component Value

Theta, θ

Q11 — Q12 — Q14 — Q22 — Q24 — Q44

Figure 5.2, Qbar Components vs. Theta

Chapter 5
Exercises

5.1. For a unidirectional ply with homogeneous transversely isotropic material properties as given below, plot the value of each $[\overline{Q}]$ component as a function of θ from $-90°$ to $90°$, as shown in figure 5.2.

Unidirectional ply material properties
$E_1 = 22.0e6$ psi
$E_2 = 1.30e6$ psi
$v_{12} = 0.30$
$v_{23} = 0.26$
$G_{12} = 0.75e6$ psi
$G_{23} = 0.516e6$ psi
$\alpha_1 = -0.30e-6$ /°F
$\alpha_2 = 18.0e-6$ /°F

Chapter 6
Classical Lamination Theory

The objective of this chapter is to derive the ABD matrices for a laminated plate using the Classical Lamination Theory (CLT). Classical Lamination Theory allows for the determination of stress and strain within each ply of a laminated plate given the force and moment resultants applied to the plate.

6.1 Laminate Conventions

In order to discuss laminated composites, we must first define the laminate stacking sequence and ply z-coordinate conventions used in this book. Note that the laminate conventions used in this book are not necessarily consistent with those used in all FEA solvers. It is recommended that users consult the specific solver documentation for the laminate conventions used within that solver.

The global coordinate system of a homogeneous or laminated plate is defined in figure 6.1. Note that the xy-plane defined by the global coordinate system goes through the middle surface of the plate with the z-axis "down" as defined using right hand rule for the xy-plane, as shown in figure 6.1. Assuming a laminated plate, the laminate stacking sequence and ply z-coordinate conventions are also defined in figure 6.1. Plies are numbered 1 through n with the 1st ply defined as the most negative z ply and the nth ply as the most positive z ply. Plies stack in the positive z direction from ply 1 to ply n. In addition, the z-coordinate value for the kth ply is always defined as the most positive z-coordinate interface for that ply. The z-coordinate values of a ply are measured relative to the middle surface of the plate. Therefore, ply 1 through ply $(n/2) - 1$ will have negative z-coordinate values. Likewise, ply $(n/2)$ through ply n will have positive z-coordinate values. Since, by definition, there will always be n + 1 ply interfaces, the Z_0 coordinate values is defined as $-T/2$.

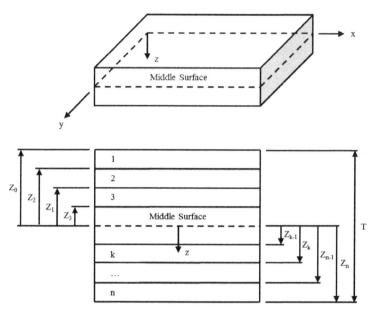

Figure 6.1, Laminated Plate Stacking Sequence and Z-coordinates

6.2 Kirchhoff Plate Theory

The Kirchhoff plate theory defines the deformation of thin plates as given by equations (6.1) and (6.2). Furthermore, Kirchhoff plate theory assumes thin plates to be under a state of plane strain. A state of plane strain is defined as $\varepsilon_z = \gamma_{xz} = \gamma_{yz} = 0$.

$$u(x,y,z) = u_o(x,y) - z\frac{\partial w_o(x,y)}{\partial x} \tag{6.1}$$

$$v(x,y,z) = v_o(x,y) - z\frac{\partial w_o(x,y)}{\partial y} \tag{6.2}$$

- $u(x,y,z)$ is the displacement field in the x-axis direction at any point within the plate.
- $v(x,y,z)$ is the displacement field in the y-axis direction at any point within the plate.
- $u_o(x,y)$ is the displacement field in the x-axis direction at any point on the middle surface of the plate.
- $v_o(x,y)$ is the displacement field in the y-axis direction at any point on the middle surface of the plate.
- $w_o(x,y)$ is the displacement field in the z-axis direction at any point on the middle surface of the plate.

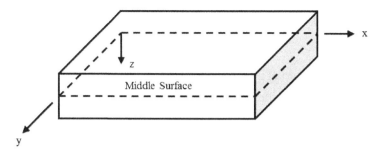

Figure 6.2, Kirchhoff Plate Definitions

The strain-displacement relationship for small-strain linear elasticity under a plane strain condition is given as;

$$\varepsilon_x = \frac{\partial u}{\partial x} \tag{6.3}$$

$$\varepsilon_y = \frac{\partial v}{\partial y} \tag{6.4}$$

$$\gamma_{xy} = \frac{\partial u}{\partial y} + \frac{\partial v}{\partial x} \tag{6.5}$$

Evaluating equations (6.3) through (6.5) using the Kirchhoff plate theory yields the total strain of a Kirchhoff plate under plane strain;

$$\varepsilon_x = \frac{\partial u_o}{\partial x} - z\frac{\partial^2 w_o}{\partial x^2} \tag{6.6}$$

$$\varepsilon_y = \frac{\partial v_o}{\partial y} - z\frac{\partial^2 w_o}{\partial y^2} \tag{6.7}$$

$$\gamma_{xy} = \frac{\partial u_o}{\partial y} + \frac{\partial v_o}{\partial x} - 2z\frac{\partial^2 w_o}{\partial x \partial y} \tag{6.8}$$

You can write the total strain of a Kirchhoff plate in matrix form as equation (6.9). Furthermore, the total strain at any z-location within the plate can further be simplified by writing equation (6.9) as equation (6.10).

$$\begin{Bmatrix} \varepsilon_x \\ \varepsilon_y \\ \gamma_{xy} \end{Bmatrix} = \begin{Bmatrix} \varepsilon_x^o \\ \varepsilon_y^o \\ \gamma_{xy}^o \end{Bmatrix} + z \begin{Bmatrix} \kappa_x \\ \kappa_y \\ \kappa_{xy} \end{Bmatrix} \tag{6.9}$$

$$\{\varepsilon\}_x = \{\varepsilon^o\}_x + z\{\kappa\}_x \qquad (6.10)$$

Where;
- $\{\varepsilon\}_x$ is the total strain at a given z-coordinate value in the plate.
- $\{\varepsilon^o\}_x$ is the middle surface strain of the plate.
- $\{\kappa\}_x$ is the middle surface curvature of the plate.

Furthermore, if the plate is regarded as a laminated plate, the total strain within the kth ply of a laminated plate can be written as;

$$\{\varepsilon\}_{x,k} = \{\varepsilon^o\}_x + z_k \{\kappa\}_x \qquad (6.11)$$

Where;
- $\{\varepsilon\}_{x,k}$ is the total strain in the global coordinate system within the kth ply.
- z_k is a valid z-coordinate value within the kth ply.

6.3 Plate Resultant Forces and Moments

The positive sign conventions for the resultant forces of a homogeneous or laminated plate are given in figure 6.3. Plate resultant forces have units of force per unit length of the plate. For a homogeneous plate of constant thickness, the resultant forces can be written in terms of stress variation through the thickness of the plate as defined by equation (6.12).

$$\begin{Bmatrix} N_x \\ N_y \\ N_{xy} \end{Bmatrix} = \int_{-t/2}^{t/2} \begin{Bmatrix} \sigma_x \\ \sigma_y \\ \tau_{xy} \end{Bmatrix} dz \qquad (6.12)$$

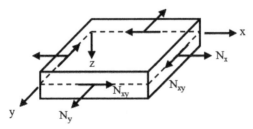

Figure 6.3, Homogeneous Plate Resultant Force Sign Convention

For a laminated plate made up of n plies, the resultant forces can be written in terms of the sum of the stress variation through the thickness of each ply as defined by equation (6.13).

$$
\begin{Bmatrix} N_x \\ N_y \\ N_{xy} \end{Bmatrix} = \sum_{k=1}^{n} \int_{z_{k-1}}^{z_k} \begin{Bmatrix} \sigma_x \\ \sigma_y \\ \tau_{xy} \end{Bmatrix} dz
\tag{6.13}
$$

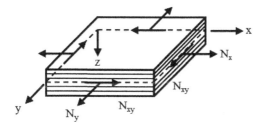

Figure 6.4, Laminated Plate Resultant Force Sign Convention

The positive sign conventions for the resultant moments of a homogeneous or laminated plate are given in figure 6.4. Plate resultant moments have units of moment (force times distance) per unit length of the plate. For a homogeneous plate of constant thickness, the resultant moments can be written in terms of stress variation through the thickness of the plate as defined by equation (6.14).

$$
\begin{Bmatrix} M_x \\ M_y \\ M_{xy} \end{Bmatrix} = \int_{-t/2}^{t/2} \begin{Bmatrix} \sigma_x \\ \sigma_y \\ \tau_{xy} \end{Bmatrix} z\,dz
\tag{6.14}
$$

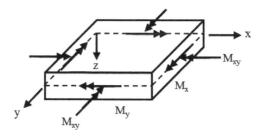

Figure 6.5, Homogeneous Plate Resultant Moment Sign Convention

For a laminated plate made up of n plies, the resultant moments can be written in terms of the sum of the stress variation through the thickness of each ply as defined by equation (6.15).

$$\begin{Bmatrix} M_x \\ M_y \\ M_{xy} \end{Bmatrix} = \sum_{k=1}^{n} \int_{z_{k-1}}^{z_k} \begin{Bmatrix} \sigma_x \\ \sigma_y \\ \tau_{xy} \end{Bmatrix} z\,dz \qquad (6.15)$$

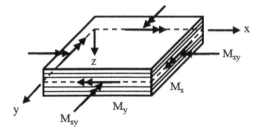

Figure 6.6, Laminated Plate Resultant Moment Sign Convention

The plate resultant forces and moments, given by equations (6.13) and (6.15), will be utilized in the next section to derive the ABD matrices for a laminated plate. The ABD matrices relate the plate resultant forces and moments to the plate middle surface strains and curvatures.

6.4 ABD Matrix of a Laminated Plate

Using equation (5.7) and adding the subscript k to designate the kth ply of a given laminate, the stress-strain relationship of the kth ply in global coordinate system is written as;

$$\{\sigma\}_{x,k} = [\overline{Q}]_k (\{\varepsilon\}_{x,k} - \{\alpha\}_{x,k} \Delta T) \qquad (6.16)$$

Substituting equations (6.11) and (6.16) into equation (6.13), one obtains the relationship between plate resultant forces and plate middle surface strains and curvatures as;

$$\{N\}_x = \sum_{k=1}^{n} \int_{z_{k-1}}^{z_k} [\overline{Q}]_k (\{\varepsilon^o\}_x + z_k \{\kappa\}_x - \{\alpha\}_{x,k} \Delta T) dz \qquad (6.17)$$

Equation (6.17) can be evaluated and rewritten as;

$$\{N\}_x = [A]\{\varepsilon^o\}_x + [B]\{\kappa\}_x - \Delta T\{N_t\}_x \qquad (6.18)$$

Where;

$$[A] = \sum_{k=1}^{n} ([\overline{Q}]_k (z_k - z_{k-1})) \qquad (6.19)$$

$$[B] = \frac{1}{2}\sum_{k=1}^{n} ([\overline{Q}]_k (z_k^2 - z_{k-1}^2)) \qquad (6.20)$$

$$\{N_t\}_x = \sum_{k=1}^{n} ([\overline{Q}]_k \{\alpha\}_{x,k} (z_k - z_{k-1})) \qquad (6.21)$$

Performing similar operations on the plate resultant moments, by substituting equations (6.11) and (6.16) into equation (6.15), one obtains the relationship between plate resultant moments and plate middle surface strains and curvatures as;

$$\{M\}_x = \sum_{k=1}^{n} \int_{z_{k-1}}^{z_k} [\overline{Q}]_k (\{\varepsilon^o\}_x + z_k \{\kappa\}_x - \{\alpha\}_{x,k} \Delta T) z \, dz \qquad (6.22)$$

Equation (6.22) can be evaluated and rewritten as;

$$\{M\}_x = [B]\{\varepsilon^o\}_x + [D]\{\kappa\}_x - \Delta T\{M_t\}_x \qquad (6.23)$$

Where;

$$[B] = \frac{1}{2}\sum_{k=1}^{n} ([\overline{Q}]_k (z_k^2 - z_{k-1}^2)) \qquad (6.24)$$

$$[D] = \frac{1}{3}\sum_{k=1}^{n} ([\overline{Q}]_k (z_k^3 - z_{k-1}^3)) \qquad (6.25)$$

$$\{M_t\}_x = \frac{1}{2}\sum_{k=1}^{n} ([\overline{Q}]_k \{\alpha\}_{x,k} (z_k^2 - z_{k-1}^2)) \qquad (6.26)$$

Combining equations (6.18) and (6.23) and writing in long form, one obtains the Classical Lamination Theory (CLT) plate equation (6.27) relating plate resultant forces and moments to middle surface strains and curvatures. $\{N_t\}_x$ and $\{M_t\}_x$ are the equivalent thermal plate resultant forces and moments in the global coordinate system.

$$\left\{ \begin{array}{c} N \\ M \end{array} \right\}_x = \left[\begin{array}{cc} A & B \\ B & D \end{array} \right] \left\{ \begin{array}{c} \varepsilon^o \\ \kappa \end{array} \right\}_x - \Delta T \left\{ \begin{array}{c} N_t \\ M_t \end{array} \right\}_x$$

$$\left\{ \begin{array}{c} N_x \\ N_y \\ N_{xy} \\ M_x \\ M_y \\ M_{xy} \end{array} \right\} = \left[\begin{array}{cccccc} A_{11} & A_{12} & A_{14} & B_{11} & B_{12} & B_{14} \\ A_{12} & A_{22} & A_{24} & B_{12} & B_{22} & B_{24} \\ A_{14} & A_{24} & A_{44} & B_{14} & B_{24} & B_{44} \\ B_{11} & B_{12} & B_{14} & D_{11} & D_{12} & D_{14} \\ B_{12} & B_{22} & B_{24} & D_{12} & D_{22} & D_{24} \\ B_{14} & B_{24} & B_{44} & D_{14} & D_{24} & D_{44} \end{array} \right] \left\{ \begin{array}{c} \varepsilon_x^o \\ \varepsilon_y^o \\ \gamma_{xy}^o \\ \kappa_x \\ \kappa_y \\ \kappa_{xy} \end{array} \right\} - \Delta T \left\{ \begin{array}{c} N_{t,x} \\ N_{t,y} \\ N_{t,xy} \\ M_{t,x} \\ M_{t,y} \\ M_{t,xy} \end{array} \right\} \quad (6.27)$$

Studying the Classical Lamination Theory (CLT) plate equation (6.27), we can make the following observations about the various components of the Classical Lamination Theory which define laminated plate behavior under any combination of applied resultant forces and moments.

The [A] matrix relates plate resultant forces to middle surface strains defining the extensional behavior of a laminate. Special notice should be given to A_{14} and A_{24} terms. These terms represent extension-shear coupling between the resultant forces and middle surface shear strain. The [A] matrix is stacking sequence independent, as will be shown in chapter 8.

The [B] matrix relates plate resultant forces to middle surface curvatures and plate resultant moments to middle surface strains. These relationships represent extension-bending, extension-twist, shear-bending, and shear-twist coupling. The [B] matrix is zero for symmetric laminates, as will be shown in chapter 8.

The [D] matrix relates plate resultant moments to middle surface curvatures defining the bending behavior of a laminate. Special notice should be given to D_{14} and D_{24} terms. These terms represent bending-twist coupling between the plate resultant moments and middle surface curvatures. The [D] matrix is stacking sequence dependent and is most affected by the location of zero degree plies in the stacking sequence, as will be shown in chapter 8.

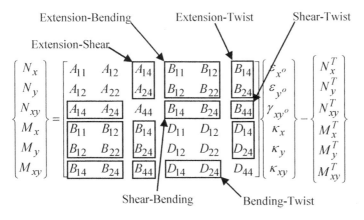

Figure 6.7, Coupling Terms of the ABD Matrix

Typically the middle surface strains and curvatures are to be determined given known resultant forces and moments. As such, equation (6.27) needs to be inverted to obtain equation (6.28). The inverse of the ABD matrix is known as the abd matrix as given in equation (6.28). In order to obtain the abd matrix, the entire ABD (6x6) matrix needs to be inverted to obtain the abd (6x6) matrix. Inverting each individual A, B, and D (3x3) matrix independently does not produce the abd (6x6) matrix as a whole. It is only possible to invert the A matrix, to obtain the a matrix, when the B matrix is zero (i.e. symmetric laminates). Similar, it is only possible to invert the D matrix to obtain the d matrix when the B matrix is zero.

$$\left\{ \begin{array}{c} \varepsilon^o \\ \kappa \end{array} \right\}_x = \begin{bmatrix} a & b \\ b & d \end{bmatrix} \left(\left\{ \begin{array}{c} N \\ M \end{array} \right\}_x + \Delta T \left\{ \begin{array}{c} N_t \\ M_t \end{array} \right\}_x \right)$$

$$\left\{ \begin{array}{c} \varepsilon^o_x \\ \varepsilon^o_y \\ \gamma^o_{xy} \\ \kappa_x \\ \kappa_y \\ \kappa_{xy} \end{array} \right\} = \begin{bmatrix} a_{11} & a_{12} & a_{14} & b_{11} & b_{12} & b_{14} \\ a_{12} & a_{22} & a_{24} & b_{12} & b_{22} & b_{24} \\ a_{14} & a_{24} & a_{44} & b_{14} & b_{24} & b_{44} \\ b_{11} & b_{12} & b_{14} & d_{11} & d_{12} & d_{14} \\ b_{12} & b_{22} & b_{24} & d_{12} & d_{22} & d_{24} \\ b_{14} & b_{24} & b_{44} & d_{14} & d_{24} & d_{44} \end{bmatrix} \left(\left\{ \begin{array}{c} N_x \\ N_y \\ N_{xy} \\ M_x \\ M_y \\ M_{xy} \end{array} \right\} + \Delta T \left\{ \begin{array}{c} N_{t,x} \\ N_{t,y} \\ N_{t,xy} \\ M_{t,x} \\ M_{t,y} \\ M_{t,xy} \end{array} \right\} \right) \quad (6.28)$$

6.6 Sequence for Solving Laminated Plate Problems

The following section describes the sequence of calculations that must be performed to solve laminated plate problems loading including thermal environments using the Classical Lamination Theory.

1. For each kth ply, calculate the plane stress compliance matrix $[S^*]_k$.
   Anisotropic Ply   (3.33)
   Isotropic Ply   (3.48)
   Transversely Isotropic Ply   (3.66)
   Orthotropic Ply   (3.87)

2. For each kth ply, calculate the plane stress stiffness matrix $[Q]_k$.
   Anisotropic Ply   (3.37)
   Isotropic Ply   (3.53)
   Transversely Isotropic Ply   (3.71)
   Orthotropic Ply   (3.92)

3. For each kth ply, calculate the plane stress stiffness matrix in the global coordinate system $[\overline{Q}]_k$.   (5.8)

4. For each kth ply, calculate the coefficient of thermal expansion in the global coordinate system $\{\alpha\}_{x,k}$ given the coefficient of thermal expansion in the material coordinate system $\{\alpha\}_{1,k}$. Note that $\{\alpha\}$ is strain per unit temperature, therefore $\{\alpha\}$ transforms as a 2nd order tensor.   (5.2)

5. Calculate the ABD matrices and the equivalent thermal resultant force and moment vectors.
   [A] matrix   (6.19)
   [B] matrix   (6.20)
   [D] matrix   (6.25)
   $\{N_t\}$ equivalent thermal resultant force   (6.21)
   $\{M_t\}$ equivalent thermal resultant moment   (6.26)

6. Calculate the inverse the ABD matrix.

7. Calculate the laminate middle surface strains $\{\varepsilon^o\}_x$ and curvatures $\{\kappa\}_x$ in the global coordinate system given the plate resultant forces and moments $\{N\}_x$, $\{M\}_x$, $\{N_t\}_x$, and $\{M_t\}_x$ in the global coordinate system.   (6.28)

8. For each kth ply, calculate the homogeneous total strain tensor in
   the global coordinate system $\{\varepsilon\}_{x,k}$.   (6.11)

9. For each kth ply, transform the homogeneous total strain tensor in
   the global coordinate system $\{\varepsilon\}_{x,k}$, into the material coordinate
   system $\{\varepsilon\}_{1,k}$.   (5.4)

10. For each kth ply, calculate the homogeneous mechanical strain
    tensor in the principal material coordinate system $\{\varepsilon\}_{m,1,k}$. (3.101)

11. For each kth ply, calculate the homogeneous principal strains $\varepsilon_{1,k}$
    and $\varepsilon_{3,k}$, and strain invariants $J_{1,k}$ and $\varepsilon_{vm,k}$.

    $\varepsilon_{1,k}$ (2.39)
    $\varepsilon_{2,k}$ (2.39)
    $J_{1,k}$ (2.38)
    $\varepsilon_{vm,k}$ (2.46)

12. For each kth ply, calculate the homogeneous stress tensor in the
    material coordinate system $\{\sigma\}_{1,k}$.

    Anisotropic   (3.37)
    Isotropic Ply   (3.51)
    Transversely Isotropic Ply   (3.69)
    Orthotropic Ply   (3.90)

13. For each kth ply, calculate the homogeneous stress tensor in the
    global coordinate system $\{\sigma\}_{x,k}$.   (5.1)

14. For each kth ply, calculate the kth ply homogeneous principal
    stresses $\sigma_{1,k}$ and $\sigma_{2,k}$, and stress invariants $\sigma_{vm,k}$.

    $\sigma_{1,k}$ (2.8)
    $\sigma_{2,k}$ (2.8)
    $\sigma_{vm,k}$ (2.21)

Chapter 6
Exercises

6.1. Calculate the [A], [B], and [D] matrices, the equivalent thermal force vector $\{N_t\}_x$, and the equivalent thermal moment vector $\{M_t\}_x$ for a $[-45/0/45/90]_s$ laminate made from unidirectional plies of thickness $t = 0.01$" with transversely isotropic material properties given below. Assume a $\Delta T = 1°F$ (i.e. a unit equivalent thermal force/moment vector)

Ply material properties
$E_1 = 22.0e6$ psi
$E_2 = 1.30e6$ psi
$v_{12} = 0.30$
$v_{23} = 0.26$
$G_{12} = 0.75e6$ psi
$G_{23} = 0.516e6$ psi
$\alpha_1 = -0.30e{-6}/°F$
$\alpha_2 = 18.0e{-6}/°F$

6.2. For the laminate in problem 6.1, calculate the inverse of the ABD matrix.

6.3. For the laminate in problem 6.1, under the boundary condition $N_x = 2400$lbs/in, calculate the laminate middle surface strains $\{\varepsilon°\}_x$ and curvatures $\{\kappa\}_x$.

6.4. For the 90° ply of the laminate in problem 6.1, under the boundary condition $N_x = 2400$lbs/in, calculate;

a) The homogeneous total strain tensor in the global system $\{\varepsilon\}_x$.
b) The homogeneous total strain tensor in the material system $\{\varepsilon\}_1$.
c) The homogeneous mech strain tensor in the material system $\{\varepsilon\}_{m,1}$.
d) The homogeneous stress tensor in the material system $\{\sigma\}_1$.

6.5. For the laminate in problem 6.1, under the boundary condition $\Delta T = -100°F$, calculate the laminate middle surface strains $\{\varepsilon^o\}_x$ and curvatures $\{\kappa\}_x$.

6.6. For the 90° ply of the laminate in problem 6.1, under the boundary condition $\Delta T = -100°F$, calculate;

a) The homogeneous total strain tensor in the global system $\{\varepsilon\}_x$.
b) The homogeneous total strain tensor in the material system $\{\varepsilon\}_1$.
c) The homogeneous mech strain tensor in the material system $\{\varepsilon\}_{m,1}$.
d) The homogeneous stress tensor in the material system $\{\sigma\}_1$.

Would you expect the 90° ply in the 1-axis fiber direction to be under tension or compression? Would you expect the 90° ply in the 2-axis transverse matrix direction to be under tension or compression? Explain your answer.

6.7. For the 90° ply of the laminate in problem 6.1, under the combined boundary condition $N_x = 2400 lbs/in$ and $\Delta T = -100°F$, calculate;

a) The homogeneous total strain tensor in the global system $\{\varepsilon\}_x$.
b) The homogeneous total strain tensor in the material system $\{\varepsilon\}_1$.
c) The homogeneous mech strain tensor in the material system $\{\varepsilon\}_{m,1}$.
d) The homogeneous stress tensor in the material system $\{\sigma\}_1$.

Chapter 7
Equivalent Homogeneous Plate Calculations

The objective of equivalent homogeneous plate calculations is to enable the mathematical modeling of complex heterogeneous laminated plates as simplified equivalent homogeneous plates. Laminated plates are by definition heterogeneous, since they are produced by layering multiple homogeneous plies of the same or different ply materials; therefore, laminated plates are not the same throughout. Even a laminated plate made with plies of all the same ply material but different fiber orientations is heterogeneous; since, each ply's fiber orientation effectively produces another material when the plane strain stiffness matrix in the material coordinate system [Q] is transformed into the global coordinate system producing the $[\overline{Q}]$ matrix. A homogeneous plate can be defined as laminated plate made up of a single homogeneous ply. The ultimate goal of equivalent homogeneous plate calculations is to mathematically equate the behavior of a laminated plate to that of a homogeneous plate, under certain limited boundary conditions, such that the behaviors of the two plates will mathematically be the same. This procedure allows for laminated plates to be simplified to homogeneous plates maintaining the same mathematical behavior without the modeling complexity. However, as will be observed within this chapter, homogeneous plates do not exhibit certain coupling behaviors that laminate plates can exhibit; therefore, equivalent homogeneous plate calculations are only valid under certain limited boundary conditions as given in each section below.

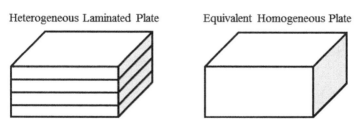

Figure 7.1, Equivalent Homogeneous Plate

7.1 ABD Matrix of a Homogeneous Plate

Extending the ABD matrix of a laminated plate, as defined in chapter 6 using the Classical Laminate Theory, the ABD matrix of a homogeneous plate can be derived. A homogeneous plate is defined as a laminated plate with a single homogeneous ply. A homogeneous plate has a stacking sequence convention, following that of section 6.1 for laminated plates, as shown in figure 7.2 below.

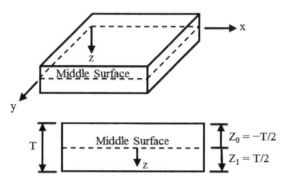

Figure 7.2, Homogeneous Plate Stacking Sequence and Z-coordinate Definitions

The ABD matrix of a laminated plate, as given in section 6.4, is repeated below for convenience.

$$[A]_{lam} = \sum_{k=1}^{n} ([\overline{Q}]_k (z_k - z_{k-1})) \tag{7.1}$$

$$[B]_{lam} = \frac{1}{2} \sum_{k=1}^{n} ([\overline{Q}]_k (z_k^2 - z_{k-1}^2)) \tag{7.2}$$

$$[D]_{lam} = \frac{1}{3} \sum_{k=1}^{n} ([\overline{Q}]_k (z_k^3 - z_{k-1}^3)) \tag{7.3}$$

$$\begin{Bmatrix} N \\ M \end{Bmatrix}_x = \begin{bmatrix} A & B \\ B & D \end{bmatrix}_{lam} \begin{Bmatrix} \varepsilon^o \\ \kappa \end{Bmatrix}_x - \Delta T \begin{Bmatrix} N_t \\ M_t \end{Bmatrix}_x \tag{7.4}$$

Inverting equation (7.4);

$$\begin{Bmatrix} \varepsilon^o \\ \kappa \end{Bmatrix}_x = \begin{bmatrix} a & b \\ b & d \end{bmatrix}_{lam} \left(\begin{Bmatrix} N \\ M \end{Bmatrix}_x + \Delta T \begin{Bmatrix} N_t \\ M_t \end{Bmatrix}_x \right) \tag{7.5}$$

The ABD matrix for a homogeneous plate is then derived from the ABD matrix of a laminated plate by considering a single homogeneous ply ($n = 1$) and evaluating $z_k = z_1 = T / 2$ and $z_{k-1} = z_0 = -T / 2$ for the single homogeneous ply as defined in figure 7.2. The resulting ABD matrix for a homogeneous plate is given by equations (7.6) through (7.8).

$$[A]_{hom} = [\overline{Q}] \, T \tag{7.6}$$

$$[B]_{hom} = 0 \tag{7.7}$$

$$[D]_{hom} = [\overline{Q}] \frac{T^3}{12} = [A]_{hom} \frac{T^2}{12} \tag{7.8}$$

T is the total thickness of the homogeneous plate as given by figure 7.2. $[\overline{Q}]$ is the plane stress stiffness matrix in global coordinates for the homogeneous plate as defined by the appropriate linear elastic material law of Chapter 3. For an isotropic material, the plane stress stiffness matrix in global coordinates from Chapter 3 is repeated below as equation (7.9).

$$[\overline{Q}]_{iso} = \begin{bmatrix} \dfrac{E}{1-v^2} & \dfrac{vE}{1-v^2} & 0 \\ \dfrac{vE}{1-v^2} & \dfrac{E}{1-v^2} & 0 \\ 0 & 0 & G \end{bmatrix} \tag{7.9}$$

It is observed from equations (7.6) through (7.8) that a homogeneous plate will always have a [B] matrix equal to zero. Therefore, a homogeneous plate will never exhibit extension-bending, extension-twist, shear-bending, or shear-twist coupling. In addition, it is observed that the [D] matrix of a homogeneous plate can be calculated from the [A] matrix of the homogeneous plate. Therefore, only calculation of the [A] matrix is required to completely define the behavior of a homogeneous plate.

For a homogeneous plate, the resultant forces and moments are related to the middle surface strains and curvatures by equations (7.10) and (7.11). Notice that since the [B] matrix of a homogeneous plate is always zero, there is no coupling of the in-plane behavior and the bending behavior of a homogeneous plate.

$$\{N\}_x = [\overline{Q}] \ T\{\varepsilon^o\}_x - \Delta T\{N_t\}_x \tag{7.10}$$

$$\{M\}_x = [\overline{Q}]\frac{T^3}{12}\{\kappa\}_x$$

Inverting equation (7.10);

$$\{\varepsilon^o\}_x = [\overline{S}^*]\frac{1}{T}(\{N\}_x + \Delta T\{N_t\}_x) \tag{7.11}$$

$$\{\kappa\}_x = [\overline{S}^*]\frac{12}{T^3}\{M\}_x$$

$[\overline{S}^*]$ is the plane stress compliance matrix in the global coordinate system for the homogeneous plate as defined by the appropriate linear elastic material law of Chapter 3. For an isotropic material, the plane stress compliance matrix in global coordinates from Chapter 3 is repeated below as equation (7.12).

$$[\overline{S}^*]_{iso} = \begin{bmatrix} \dfrac{1}{E} & \dfrac{-\nu}{E} & 0 \\ \dfrac{-\nu}{E} & \dfrac{1}{E} & 0 \\ 0 & 0 & \dfrac{1}{G} \end{bmatrix} \tag{7.12}$$

7.2 Equivalent In-Plane Homogeneous Engineering Constants

For a given laminated plate, the equivalent in-plane homogenized engineering constants in the global coordinate system (E_x, E_y, G_{xy}, and ν_{xy}) can be determined by considering a plate under the boundary conditions, $\{M\}_x = 0$ and $\Delta T = 0$. In addition, there is a requirement that the given laminated plate stacking sequence be balanced and symmetric (i.e. $A_{14} = A_{24} = 0$ and $[B] = 0$). Under these conditions, the in-plane behavior of a laminated plate is decoupled from its bending behavior. Furthermore, the laminated plate resultant forces are related to the laminated plate middle surface strains by equation (7.13). Also under these conditions, the homogeneous plate forces are related to the homogeneous plate middle surface strains by equation (7.14).

$$\{\varepsilon^o\}_x = [a]_{lam}\{N\}_x \tag{7.13}$$

$$\{\varepsilon^o\}_x = [\bar{S}*]\frac{1}{T}\{N\}_x \tag{7.14}$$

Equating the laminate and homogeneous plate middle surface strains given by equations (7.13) and (7.14) respectively, equation (7.15) can be written which relates the laminated plate abd matrix to the homogeneous plate plane stress compliance matrix in the global coordinate system. Using the transversely isotropic linear elastic material law, the homogeneous plane stress compliance matrix in the global coordinate system can be written as given by equation (7.16).

$$[a]_{lam} = [\bar{S}*]\frac{1}{T} \tag{7.15}$$

$$\begin{bmatrix} a_{11} & a_{12} & 0 \\ a_{12} & a_{22} & 0 \\ 0 & 0 & a_{44} \end{bmatrix}_{lam} = \begin{bmatrix} \dfrac{1}{E_x T} & \dfrac{-v_{xy}}{E_x T} & 0 \\ \dfrac{-v_{xy}}{E_x T} & \dfrac{1}{E_y T} & 0 \\ 0 & 0 & \dfrac{1}{G_{xy} T} \end{bmatrix} \tag{7.16}$$

Equating each component of equation (7.16), the equivalent in-plane homogenized engineering constants are derived as given by equation (7.17).

$$E_x = \frac{1}{T\, a_{11,lam}}$$

$$E_y = \frac{1}{T\, a_{22,lam}} \tag{7.17}$$

$$v_{xy} = \frac{-a_{12,lam}}{a_{11,lam}}$$

$$G_{xy} = \frac{1}{T\, a_{44,lam}}$$

Where;

$$[a]_{lam} = [A]_{lam}^{-1}$$

It is important to recognize that the equivalent in-plane homogenized engineering constants do not account for any coupling behavior. Therefore, the equivalent in-plane homogenized engineering constants, calculated using equation (7.17), are only valid for laminates which do not exhibit coupling behaviors. Laminates which do not exhibit coupling behaviors must be balanced and symmetric to adhere to these requirements. In addition, the equivalent in-plane homogenized engineering constants are only applicable to plates dominated by resultant forces. Equation (7.17) cannot be utilized for plates that are dominated by resultant moments. In the case a plate is dominated by resultant moments, the equivalent homogenized engineering constants should be derived from the [D] matrix as is derived in section 7.3.

7.3 Equivalent Bending Homogeneous Engineering Constants

For a given laminated plate, the equivalent bending homogenized engineering constants in the global coordinate system (E_x, E_y, G_{xy}, and v_{xy}) can be determined by considering a plate under the boundary conditions, $\{N\}_x = 0$ and $\Delta T = 0$. In addition, there is a requirement that the given laminated plate stacking sequence be symmetric (i.e. [B] = 0). Under these conditions, the bending behavior of a laminated plate is decoupled from its in-plane behavior. Furthermore, the laminated plate resultant moments are related to the laminated plate middle surface curvatures by equation (7.18). Also under these conditions, the homogeneous plate forces are related to the homogeneous plate middle surface strains by equation (7.19).

$$\{\kappa\}_x = [d]_{lam}\{M\}_x \tag{7.18}$$

$$\{\kappa\}_x = [\bar{S}^*]\frac{12}{T^3}\{M\}_x \tag{7.19}$$

Equating the laminate and homogeneous plate middle surface curvatures given by equations (7.18) and (7.19) respectively, equation (7.20) can be written which relates the laminated plate abd matrix to the homogeneous plate plane stress compliance matrix in the global coordinate system. Using the transversely isotropic linear elastic material law, the homogeneous plane stress compliance matrix in the global coordinate system can be written as given by equation (7.21).

$$[d]_{lam} = [\bar{S}*]\frac{12}{T^3}$$
(7.20)

$$\begin{bmatrix} d_{11} & d_{12} & d_{14} \\ d_{12} & d_{22} & d_{24} \\ d_{14} & d_{24} & d_{44} \end{bmatrix}_{lam} = \begin{bmatrix} \dfrac{12}{E_x T^3} & \dfrac{-12\nu_{xy}}{E_x T^3} & 0 \\ \dfrac{-12\nu_{xy}}{E_x T} & \dfrac{12}{E_y T} & 0 \\ 0 & 0 & \dfrac{12}{G_{xy} T} \end{bmatrix}$$
(7.21)

Equating each component of equation (7.21), the equivalent bending homogenized engineering constants are derived as given by equation (7.22).

$$E_x = \frac{12}{T^3 d_{11,lam}}$$

$$E_y = \frac{12}{T^3 d_{22,lam}}$$
(7.22)

$$\nu_{xy} = \frac{-d_{12,lam}}{d_{11,lam}}$$

$$G_{xy} = \frac{12}{T^3 d_{44,lam}}$$

Where;
$$[d]_{lam} = [D]_{lam}^{-1}$$

It is important to recognize that the equivalent bending homogenized engineering constants do not account for any coupling behavior. Therefore, the equivalent bending homogenized engineering constants, calculated using equation (7.20), are only valid for laminates that do not exhibit coupling behaviors. Laminates that do not exhibit coupling behaviors must be symmetric to adhere to these requirements. In addition, the equivalent bending homogenized engineering constants are only applicable to plates dominated by resultant moments. Equation (7.22) cannot be utilized for plates that are dominated by resultant forces. In the case a plate is dominated by resultant forces, the equivalent homogenized engineering constants should be derived from the [A] matrix as is derived in section 7.2.

7.4 Equivalent Homogeneous Material Matrices

Most modern finite element solvers have anisotropic homogenized plate element formulations that can exhibit all forms of extension-bending coupling behaviors. Anisotropic homogenized plate element formulations allow for homogenized plates to exactly behave as laminated plates under all conditions. However, since the laminated plate stacking sequence is not available to the homogenized plate, detailed ply-level stress and strain output is not available for the homogenized plate elements. In order for these homogenized plate elements to exhibit extension-bending coupling, most modern finite element solvers write the relationship between the plate resultant forces and moments and the middle surface strains and curvatures for a homogenized plate as given by equation (7.23).

$$
\left\{ \begin{array}{c} N \\ M \\ Q \end{array} \right\}_x = \left[\begin{array}{ccc} \overline{Q}_1 T & \overline{Q}_4 T^2 & 0 \\ \overline{Q}_4 T^2 & \overline{Q}_2 \dfrac{T^3}{12} & 0 \\ 0 & 0 & \overline{Q}_3 T_s \end{array} \right] \left\{ \begin{array}{c} \varepsilon^o \\ \kappa \\ \gamma \end{array} \right\}_x - \left\{ \begin{array}{c} N_t \\ M_t \\ 0 \end{array} \right\}_x
$$
(7.23)

Where;

$[\overline{Q}_1] = \dfrac{1}{T}[A]_{lam}$ defines the effective extension material matrix.   (7.24)

$[\overline{Q}_2] = \dfrac{12}{T^3}[D]_{lam}$ defines the effective bending material matrix.   (7.25)

$[\overline{Q}_4] = -\dfrac{1}{T^2}[B]_{lam}$ defines the effective coupling material matrix.   (7.26)

And;

$\{Q\}_x = \left\{ \begin{array}{c} Q_{xz} \\ Q_{yz} \end{array} \right\}$ are the resultant transverse shear forces.

$\{\gamma\}_x = \left\{ \begin{array}{c} \gamma_{xz} \\ \gamma_{yz} \end{array} \right\}$ are the middle surface transverse shear strains.

T_s is the effective transverse shear material thickness (default = 0.8333).

$[\overline{Q}_3]$ is the effective transverse shear material matrix (2x2).

For OptiStruct, the anisotropic homogeneous plate element is defined with a PSHELL card. The PSHELL card has MID1, MID2, MID3, and MID4 fields that define the extension, bending, transverse shear, and coupling matrices for an anisotropic homogenized plate respectively. Each MIDi references a

MAT2 card that defines the appropriate effective material matrix. Each effective material matrix is a (3x3) matrix, except the MID3 field MAT2 effective transverse shear material matrix, which is a (2x2) and is discussed further below. The calculation of the effective transverse shear material matrix is fairly complicated and involved, but is presented below for completeness without derivation.

$$[\overline{Q}_3] = \begin{bmatrix} \overline{S}_{3_{11}} & \overline{S}_{3_{12}} \\ \overline{S}_{3_{21}} & \overline{S}_{3_{22}} \end{bmatrix}^{-1} \tag{7.27}$$

Where;

$$\overline{Q}_{3_{12}} = \overline{Q}_{3_{21}} = avg(\overline{Q}_{3_{12}}, \overline{Q}_{3_{21}})$$

$$\overline{S}_{3_{ij}} = \frac{T}{(\overline{EI}_{ii})^2} \sum_{k=1}^{n} \overline{S}_{ij_k} R_{i_k} \tag{7.28}$$

$$\overline{EI}_{ii} = D_{ii} - 2\overline{z}_i B_{ii} + \overline{z}_i^2 A_{ii} \tag{7.29}$$

$$\overline{z}_i = \frac{B_{ii}}{A_{ii}} \tag{7.30}$$

$$[\overline{S}]_k = \begin{bmatrix} \overline{S}_{11_k} & \overline{S}_{12_k} \\ \overline{S}_{21_k} & \overline{S}_{22_k} \end{bmatrix} = \begin{bmatrix} \cos\theta_k & -\sin\theta_k \\ \sin\theta_k & \cos\theta_k \end{bmatrix} [S]_k \begin{bmatrix} \cos\theta_k & \sin\theta_k \\ -\sin\theta_k & \cos\theta_k \end{bmatrix} \tag{7.31}$$

$$[S]_k = \begin{bmatrix} \dfrac{1}{G_{13_k}} & 0 \\ 0 & \dfrac{1}{G_{23_k}} \end{bmatrix} \tag{7.32}$$

$$R_{i_k} = \overline{Q}_{ii_k}^2 t_k [(f_{i_k} + (\overline{z}_i - z_{k-1})t_k - \frac{1}{3}t_k^2)f_{i_k} + (\frac{1}{3}(\overline{z}_i - 2z_{k-1}) - \frac{1}{4}t_k)\overline{z}_i t_k^2 \tag{7.33}$$

$$+ (\frac{1}{3}z_{k-1}^2 + \frac{1}{4}z_{k-1}t_k + \frac{1}{20}t_k^2)t_k^2]$$

$$f_{i_k} = \frac{1}{\overline{Q}_{ii_k}} \sum_{m=1}^{k-1} \overline{Q}_{ii_m} t_m [\overline{z}_i - \frac{1}{2}(z_m + z_{m-1})] \tag{7.34}$$

$$f_{1_1} = f_{2_1} = 0$$

Where;
- $i = 1, 2$
- $j = 1, 2$
- $k = 1 - n$ ply.
- T is the total thickness of the plate.
- $\overline{EI}_{ii}$ is the plate moment of inertia about the x- and y-axis.
- $\overline{z}_i$ is the z-coordinate of the plate neutral axis in the x- and y-axis.
- $[\overline{S}]_k$ is the transverse shear compliance matrix in the global coordinate system for the kth ply.
- $[S]_k$ is the transverse shear compliance matrix in the material coordinate system for the kth ply.
- $\overline{Q}_{ii_k}$ is the plane stress stiffness matrix diagonal terms for the kth ply.
- t_k is the thickness of the kth ply.
- z_k is the z-coordinate of the kth ply.

The process for calculating the effective transverse shear material matrix of a laminated plate is as follows;

1. For the laminate calculate ($i = 1, 2$).

    A_{ii} (7.1)

    B_{ii} (7.2)

    D_{ii} (7.3)

    $\overline{z}_i$ (7.30)

    $\overline{EI}_{ii}$ (7.29)

2. For each kth ply of the laminate calculate ($i = 1, 2$).

    f_{i_k} (7.34)

    R_{i_k} (7.33)

    $[S]_k$ (7.32)

    $[\overline{S}]_k$ (7.31)

3. Calculate the effective transverse shear material matrix for the laminate

    $[\overline{S}_3]$ (7.28)

    $[\overline{Q}_3]$ (7.27)

7.5 Equivalent In-Plane Quasi-Isotropic Laminate

For a given homogeneous plate, the equivalent in-plane quasi-isotropic laminate can be determined by considering a plate under the boundary conditions, $\{M\}_x = 0$ and $\Delta T = 0$. In addition, there is a requirement that the given laminated plate stacking sequence be quasi-isotropic ($T_0 = T_{90}$; $A_{11} = A_{22}$), balanced ($T_{45} = T_{-15}$; $A_{14} = A_{24} = 0$), and symmetric ($[B] = 0$). Under these conditions, the in-plane behavior of a laminated plate is decoupled from its bending behavior. In addition, the laminated plate resultant forces are related to the laminated plate middle surface strains by equation (7.35), and the homogeneous plate resultant forces are related to the homogeneous plate middle surface strains by equation (7.36).

$$\{N\}_x = [A]_{quasi-iso} \{\varepsilon^0\}_x \tag{7.35}$$

$$\{N\}_x = [\overline{Q}]_{iso} T \{\varepsilon^0\}_x \tag{7.36}$$

Equating the laminate and homogeneous plate resultant forces given by equations (7.35) and (7.36) respectively, equation (7.37) can be written. This equation relates the laminated plate [A] matrix to the homogeneous plate plane stress stiffness matrix in the global coordinate system. Using the linear elastic isotropic material law, the homogeneous plate plane stress stiffness matrix in the global coordinate system can be written as given by equation (7.38).

$$[A]_{quasi-iso} = [\overline{Q}]_{iso} T \tag{7.37}$$

$$
\begin{bmatrix} A_{11} & A_{12} & 0 \\ A_{12} & A_{22} & 0 \\ 0 & 0 & A_{44} \end{bmatrix} =
\begin{bmatrix} \dfrac{ET}{1-v^2} & \dfrac{vET}{1-v^2} & 0 \\ \dfrac{vET}{1-v^2} & \dfrac{ET}{1-v^2} & 0 \\ 0 & 0 & GT \end{bmatrix}
\tag{7.38}
$$

Since the laminated plate [A] matrix is stacking sequence independent and utilizing the fact that $T_0 = T_{90}$ due to the quasi-isotropic condition and that $T_{45} = T_{-45}$ due to the balanced condition, the quasi-isotropic laminated plate [A] matrix can be calculated as given by equation (7.45). It is important to realize that T_0, T_{90}, T_{45}, and T_{-45} are the sum of all ply thicknesses of a given ply angle within the laminate as given by equations (7.39) through (7.42).

$$T_0 = \sum_{k=1}^{i} t_{0,k} \tag{7.39}$$

$$T_{90} = \sum_{k=1}^{i} t_{90,k} \tag{7.40}$$

$$T_{45} = \sum_{k=1}^{i} t_{45,k} \tag{7.41}$$

$$T_{-45} = \sum_{k=1}^{i} t_{-45,k} \tag{7.42}$$

$$T_{90} = T_0 \tag{7.43}$$

$$T_{-45} = T_{45} \tag{7.44}$$

$$[A]_{quasi-iso} = \left([\overline{Q}]_0 + [\overline{Q}]_{90} \right) T_0 + \left([\overline{Q}]_{45} + [\overline{Q}]_{-45} \right) T_{45} \tag{7.45}$$

Since there are two unknowns, T_0 and T_{45}, there must be two independent equations to solve for the two unknowns. These two equations are given by relating the 11 and 44 components of the quasi-isotropic laminated plate [A] matrix and the homogenous plate plane stress stiffness matrix given in equation (7.38) and represented below as equations (7.46) and (7.47). Therefore, the in-plane extensional and shear behavior of the equivalent quasi-isotropic laminated plate will match that of the homogeneous plate.

$$A_{11} = \frac{ET}{1-v^2} \tag{7.46}$$

$$A_{44} = GT \tag{7.47}$$

The A_{11} and A_{44} components of the quasi-isotropic laminated plate are given below in equations (7.48) and (7.49) respectively.

$$A_{11} = \left(\overline{Q}_{11,0} + \overline{Q}_{11,90} \right) T_0 + \left(\overline{Q}_{11,45} + \overline{Q}_{11,-45} \right) T_{45} \tag{7.48}$$

$$A_{44} = \left(\overline{Q}_{44,0} + \overline{Q}_{44,90} \right) T_0 + \left(\overline{Q}_{44,45} + \overline{Q}_{44,-45} \right) T_{45} \tag{7.49}$$

The equations which relate the 11 and 44 components between the quasi-isotropic laminated plate and the homogeneous plate in matrix form is given in equation (7.50).

$$\begin{bmatrix} \left(\overline{Q}_{11,0} + \overline{Q}_{11,90}\right) & \left(\overline{Q}_{11,45} + \overline{Q}_{11,-45}\right) \\ \left(\overline{Q}_{44,0} + \overline{Q}_{44,90}\right) & \left(\overline{Q}_{44,45} + \overline{Q}_{44,-45}\right) \end{bmatrix} \begin{Bmatrix} T_0 \\ T_{45} \end{Bmatrix} = \begin{Bmatrix} \dfrac{ET}{1-\nu^2} \\ GT \end{Bmatrix} \tag{7.50}$$

Inverting equation (7.50) using crammers rule, the solution for the two thicknesses, T₀ and T₄₅, of the equivalent in-plane quasi-isotropic laminated plate are given by equation (7.51).

$$\begin{Bmatrix} T_0 \\ T_{45} \end{Bmatrix} = \begin{bmatrix} \dfrac{\left(\overline{Q}_{44,45} + \overline{Q}_{44,-45}\right)}{D} & \dfrac{-\left(\overline{Q}_{11,45} + \overline{Q}_{11,-45}\right)}{D} \\ \dfrac{-\left(\overline{Q}_{44,0} + \overline{Q}_{44,90}\right)}{D} & \dfrac{\left(\overline{Q}_{11,0} + \overline{Q}_{11,90}\right)}{D} \end{bmatrix} \begin{Bmatrix} \dfrac{ET}{1-\nu^2} \\ GT \end{Bmatrix} \tag{7.51}$$

Where;

$$D = (\overline{Q}_{11,0} + \overline{Q}_{11,90})(\overline{Q}_{44,45} + \overline{Q}_{44,-45}) - (\overline{Q}_{11,45} + \overline{Q}_{11,-45})(\overline{Q}_{44,0} + \overline{Q}_{44,90})$$

Note that the total thickness of the equivalent in-plane quasi-isotopic laminated plate is $2(T_0 + T_{45})$ and that the thickness of the 90° and -45° plies is as given by equations (7.43) and (7.44) respectively.

7.6 Equivalent Bending Quasi-Isotropic Laminate

For a given homogeneous plate, the equivalent bending quasi-isotropic laminate can be determined by considering a plate under the boundary conditions, $\{N\}_x = 0$ and $\Delta T = 0$. In addition, there is a requirement that the given laminated plate stacking sequence be quasi-isotropic ($T_0 = T_{90}$), balanced ($T_{45} = T_{-45}$), and symmetric ([B] = 0). Under these conditions, the bending behavior of a laminated plate is decoupled from its in-plane behavior. In addition, the laminated plate resultant moments are related to the laminated plate middle surface curvatures by equation (7.52), and the homogeneous plate resultant moments are related to the homogeneous plate middle surface curvatures by equation (7.53).

$$\{M\}_x = [D]_{smear}\{\kappa\}_x = [A]_{quasi-iso} \frac{T_{lam}^2}{12}\{\kappa\}_x \tag{7.52}$$

$$\{M\}_x = [\overline{Q}]_{iso} \frac{T^3}{12}\{\kappa\}_x \tag{7.53}$$

Equating the laminate and homogeneous plate resultant forces given by equations (7.52) and (7.53) respectively, equation (7.54) can be written. This equation relates the laminated plate [A] matrix to the homogeneous plate plane stress stiffness matrix in the global coordinate system. Using the linear elastic isotropic material law, the homogeneous plate plane stress stiffness matrix in the global coordinate system can be written as given by equation (7.55).

$$[A]_{quasi-iso} T_{lam}^2 = [\overline{Q}]_{iso} T^3 \tag{7.54}$$

$$
\begin{bmatrix}
A_{11}T_{lam}^2 & A_{12}T_{lam}^2 & 0 \\
A_{12} & A_{22}T_{lam}^2 & 0 \\
0 & 0 & A_{44}T_{lam}^2
\end{bmatrix}
=
\begin{bmatrix}
\dfrac{ET^3}{1-v^2} & \dfrac{vET^3}{1-v^2} & 0 \\
\dfrac{vET^3}{1-v^2} & \dfrac{ET^3}{1-v^2} & 0 \\
0 & 0 & GT^3
\end{bmatrix}
\tag{7.55}
$$

Since the laminated plate [A] matrix is stacking sequence independent and utilizing the fact that $T_0 = T_{90} = T_{45} = T_{-45}$ due to the quasi-isotropic and balanced condition, the quasi-isotropic laminated plate [A] matrix can be calculated as given by equation (7.56). It is important to realize that T_0, T_{90}, T_{45}, and T_{-45} are the sum of all ply thicknesses of that given angle as given by equations (7.39) through (7.42).

$$T_{90} = T_{45} = T_{-45} = T_{90} \tag{7.56}$$

$$T_{lam} = 4T_0 \tag{7.57}$$

$$[A]_{quasi-iso} = ([\overline{Q}]_0 + [\overline{Q}]_{90} + [\overline{Q}]_{45} + [\overline{Q}]_{-45}) T_0 \tag{7.58}$$

Since there is only one unknown, T_0, there can only be one independent equation to solve. This equation is given by relating the 11 component of the quasi-isotropic laminated plate [A] matrix and the homogenous plate plane stress stiffness matrix given in equation (7.55). The A_{11} component of the quasi-isotropic laminated plate is given in equation (7.59).

$$A_{11} = (\overline{Q}_{11,0} + \overline{Q}_{11,90} + \overline{Q}_{11,45} + \overline{Q}_{11,-45})T_0 \tag{7.59}$$

The equation which relate the 11 component between the quasi-isotropic laminated plate and the homogeneous plate is given in equation (7.60).

$$\left(\overline{Q}_{11,0} + \overline{Q}_{11,90} + \overline{Q}_{11,45} + \overline{Q}_{11,-45}\right) 16T_0^3 = \frac{ET^3}{1 - v^2} \tag{7.60}$$

Solving equation (7.60) for T_0, the equivalent bending quasi-isotropic laminated plate is given by equation (7.61).

$$T_0 = \sqrt[3]{\frac{ET^3}{16(1 - v^2)(\overline{Q}_{11,0} + \overline{Q}_{11,90} + \overline{Q}_{11,45} + \overline{Q}_{11,-45})}} \tag{7.61}$$

Note that the total thickness of the equivalent bending quasi-isotopic laminated plate is $4T_0$ as given by equation (7.57) and that the thickness of the $90°$, $45°$, and $-45°$ plies is as given by equation (7.56).

Chapter 7
Exercises

7.1. Calculate the [A], [B], and [D] matrices for a homogeneous isotropic aluminum plate of thickness T = 0.08"

Aluminum material properties
E = 10e6 psi
$v = 0.33$.
G = 3.76e6 psi

7.2. For the laminate in problem 6.1, calculate the equivalent in-plane homogenized engineering constants (E_x, E_y, v_{xy}, and G_{xy}).

7.3. For the laminate in problem 6.1, calculate the equivalent bending homogenized engineering constants (E_x, E_y, v_{xy}, and G_{xy}).

7.4. For the laminate in problem 6.1, calculate the equivalent material matrices ($[\overline{Q_1}]$, $[\overline{Q_2}]$, and $[\overline{Q_4}]$).

7.5. (Optional) For the laminate in problem 6.1, calculate the equivalent transverse shear material matrix ($[\overline{Q_3}]$).

7.6. For the homogeneous isotropic aluminum plate given in problem 7.1 with T = 0.08", calculate the equivalent in-plane quasi-isotropic laminate made from unidirectional plies with material properties as given below.

Ply material properties
$E_1 = 22.0e6$ psi
$E_2 = 1.30e6$ psi
$v_{12} = 0.30$
$v_{23} = 0.26$
$G_{12} = 0.75e6$ psi
$G_{23} = 0.516e6$ psi

7.7 For the homogeneous isotropic aluminum plate given in problem 7.1 with T = 0.08", calculate the equivalent bending quasi-isotropic laminate made from unidirectional plies with material properties as given in problem 7.6.

Chapter 8
SMEAR Technology

SMEAR technology is used to calculate a stacking sequence independent ABD matrix for a laminated plate. This is accomplished by utilizing the homogeneous plate [B] and [D] matrix equations with the laminated plate [A] matrix. The laminated plate [A] matrix is stacking sequence independent, as will be shown in this chapter. Furthermore, since the [D] matrix for a homogeneous plate is also stacking sequence independent, utilization of the homogeneous plate [D] matrix with the laminated plate [A] matrix allows for the calculation of a stacking sequence independent [D] matrix. The stacking sequence independent [D] matrix is also referred to as the [D]$_{smear}$ matrix. The [D]$_{smear}$ matrix is the average [D] matrix for all possible combinations of stacking sequences for a given laminate. SMEAR technology is used in composite design optimization, where the stacking sequence of the laminate being optimized is initially indeterminate. The details of SMEAR technology and its application to composite optimization are discussed in detail in this chapter.

8.1 Stacking Sequence Dependence of [ABD] Matrix

In this section, we discuss the stacking sequence dependence of a laminated plate ABD matrix by further investigating the classical laminate equations repeated here as equations (8.1) through (8.3). First, we will study the effects of the each ply's $[\bar{Q}]$ matrix on the laminated plate ABD matrix. The $[\bar{Q}]$ matrix, as defined in Chapter 5, is the plane stress stiffness matrix in the global coordinate system for the kth ply. Second, we will study the effects of the ($z_k^n - z_{k-1}^n$) geometric term on the laminate plate ABD matrix. Finally, we will combine the effects of the two studies and make some general observations about the existence of various components of the laminated plate ABD matrix; thus, describing the expected laminate behavior. We will conclude by applying these observations to the application of SMEAR technology for the use of calculating a stacking sequence independent laminate plate ABD matrix.

The ABD matrix for a laminated plate, as derived in Section 6.4 using Classical Lamination Theory, is repeated below for convenience.

$$[A]_{lam} = \sum_{k=1}^{n} ([\overline{Q}]_k (z_k - z_{k-1})) \tag{8.1}$$

$$[B]_{lam} = \frac{1}{2} \sum_{k=1}^{n} ([\overline{Q}]_k (z_k^2 - z_{k-1}^2)) \tag{8.2}$$

$$[D]_{lam} = \frac{1}{3} \sum_{k=1}^{n} ([\overline{Q}]_k (z_k^3 - z_{k-1}^3)) \tag{8.3}$$

Figure 8.1 shows a plot the plane stress stiffness matrix in global coordinates $[\overline{Q}]$ versus the fiber orientation for the kth ply. Looking at the plot in figure 8.1, the following statements can be made about the various components of the $[\overline{Q}]$ matrix for the kth ply as the fiber orientation varies from -90° to 90°;

- $\overline{Q}_{11}$ and $\overline{Q}_{22}$ are ninety degrees out of phase. This means a 90° ply behaves as a 0° ply in the global y-axis direction and vice versa.
- $\overline{Q}_{11}$ and $\overline{Q}_{22}$ are always positive. This means that a positive extensional strain will cause a positive extensional stress.
- $\overline{Q}_{12}$ is always positive. This means that a positive extensional strain in the global y-axis direction will try to decrease the extensional strain in the global x-axis direction, which requires that additional extensional stress (i.e. positive $\overline{Q}_{12}$) be applied in the global x-axis direction to maintain the given extensional strain in the global x-axis direction.
- $\overline{Q}_{44}$ is always positive. This means that a positive shear strain will cause a positive shear stress.
- $\overline{Q}_{14}$ and $\overline{Q}_{24}$ represent extension-shear coupling terms. They are positive for positive θ and negative for negative θ. This means that a positive shear strain will cause an additional tension stress and a negative shear strain will cause a reduction in tension stress for positive θ plies. The reverse is true for negative θ plies. This makes sense, as for a positive θ ply under positive shear, the fibers need to elongate, whereas the same positive θ ply under negative shear will cause a shorting of the fibers. In addition, $\overline{Q}_{14}$ and $\overline{Q}_{24}$ components are zero for θ equal to 0 and 90. This means extension-shear coupling does not exist for 0° and 90° plies.

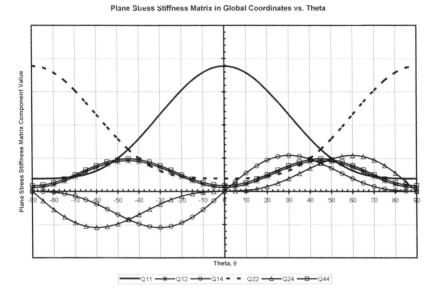

Figure 8.1, Plane Stress Stiffness Matrix in Global Coordinates vs. θ

Figure 8.2 depicts a general laminate with unit thickness plies. The z-coordinate value for each ply interface is shown on the right side of the figure according to the z-coordinate conventions given in figure 6.1. In studying the $(z_k^n - z_{k-1}^n)$ geometric terms, we will assume that each ply is made of the same homogeneous material. However, each ply can have its own unique fiber orientation θ. Table 8.1 shows the values for the $(z_k^n - z_{k-1}^n)$ geometric terms for each ply for a laminate with unit thick plies.

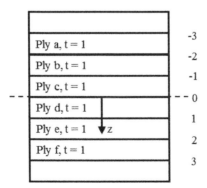

Figure 8.2, Laminate with Unit Thickness Plies

Table 8.1, $(z_k^n - z_{k-1}^n)$ Geometric Terms for Unit Thick Plies

	Z coord	[A] Matrix	[B] Matrix	[D] Matrix
	z_k	$z_k - z_{k-1}$	$z_k^2 - z_{k-1}^2$	$z_k^3 - z_{k-1}^3$
...	-3			$(-3)^3 - (-4)^3 = 37$
Ply a	-2	$(-2) - (-3) = 1$	$(-2)^2 - (-3)^2 = -5$	$(-2)^3 - (-3)^3 = 19$
Ply b	-1	$(-1) - (-2) = 1$	$(-1)^2 - (-2)^2 = -3$	$(-1)^3 - (-2)^3 = 7$
Ply c	0	$(0) - (-1) = 1$	$(0)^2 - (-1)^2 = -1$	$(0)^3 - (-1)^3 = 1$
Ply d	1	$(1) - (0) = 1$	$(-1)^2 - (0)^2 = 1$	$(1)^3 - (0)^3 = 1$
Ply e	2	$(2) - (1) = 1$	$(2)^2 - (1)^2 = 3$	$(2)^3 - (1)^3 = 7$
Ply f	3	$(3) - (2) = 1$	$(3)^2 - (2)^2 = 5$	$(3)^3 - (2)^3 = 19$
...				$(4)^3 - (3)^3 = 37$

First we will study the [A] matrix ($z_k - z_{k-1}$) geometric terms. Remember from Chapter 6 that the [A] matrix describes the extensional behavior of the laminate and relates the plate resultant forces to the middle surface strains. We should notice from table 8.1 that the [A] matrix ($z_k - z_{k-1}$) geometric terms are always positive and equal to the thickness of the ply. Therefore, the [A] matrix for a laminated plate is always stacking sequence independent, as the plies are always multiplied by their thickness independent of their location in the laminate stack. The [A] matrix for a [0/0/45/45/90/90/−45/−45] laminate is equal to the [A] matrix for a [0/45/90/−45/−45/90/45/0] laminate. Furthermore, laminates of any stacking sequence made of the same number of plies of the same fiber orientations will have same [A] matrix. In addition, since only the $\overline{Q}_{14}$ and $\overline{Q}_{24}$ terms change sign with positive and negative θ, balanced laminates will have A_{14} and A_{24} terms equal to zero; therefore, laminates with these characteristics will not exhibit extension-shear coupling.

Next we will study the [B] matrix ($z_k^2 - z_{k-1}^2$) geometric terms. Remember from Chapter 6 that the [B] matrix describes the extension-bending, extension-twist, shear-bending, and shear-twist coupling behavior of a laminated plate. The [B] matrix also relates the plate resultant forces to middle surface curvatures and plate resultant moments to middle surface strains. We should notice from table 8.1 that the [B] matrix ($z_k^2 - z_{k-1}^2$) geometric terms are negative for negative z-coordinate plies and positive for positive z-coordinate plies. In addition, the ($z_k^2 - z_{k-1}^2$) geometric terms are symmetric about the middle surface. Therefore, the [B] matrix is zero for symmetric laminates.

Finally we will study the [D] matrix ($z_k^3 - z_{k-1}^3$) geometric terms. Remember from Chapter 6 that the [D] matrix describes the bending behavior of a laminate and relates the plate resultant moments to the middle surface curvatures. We should note from table 8.1 that the [D] matrix ($z_k^3 - z_{k-1}^3$) terms are always positive and increases exponentially for plies further away from the middle surface. Therefore, the [D] matrix is stacking sequence dependent. Since $0°$ plies have the highest $\overline{Q}_{11}$ values, their location in the stacking sequence of a laminate greatly influences the bending behavior of that laminate in the longitudinal x-direction. The further away from the middle surface the $0°$ plies, the "stiffer" the bending behavior in that direction. The same holds for $90°$ plies in the transverse y-direction. If a laminate is balanced and anti-symmetric, and since only the $\overline{Q}_{14}$ and $\overline{Q}_{24}$ terms change sign with positive and negative θ, balanced and anti-symmetric laminates will have D_{14} and D_{24} terms equal to zero and those laminates will not exhibit bending-twisting coupling. In general D_{14} and D_{24} terms become insignificant for laminates with sixteen plies or greater.

8.2 SMEAR Technology in Composite Design

The objective for using SMEAR technology with composite concept design optimization is to ultimately make the concept design optimization process stacking sequence independent. This is very important, as the objective in the concept design optimization phase is to define a laminate stacking sequence when one currently is not defined. First, we will review the conclusions from section 7.1 and section 8.1.

Section 7.1 conclusions for homogeneous plates;

- Only the [A] matrix of a homogeneous plate needs to be calculated to completely define the behavior of a homogeneous plate since the [D] matrix is related to the [A] matrix for homogeneous plates.

Section 8.1 conclusions for laminated plates;

- The [A] matrix is stacking sequence independent for any laminate.
- A balanced laminate will have A_{14} and A_{24} terms equal to zero.
- The [B] matrix is zero for a symmetric laminate.
- The [D] matrix is stacking sequence dependent for any laminate.
- The [D] matrix is highly influenced by the stack location of the $0°$ plies.
- A balanced and anti-symmetric laminate will have D_{14} and D_{24} terms equal to zero.

Using these conclusions, we can make the laminated plate ABD matrix completely stacking sequence independent by using the following combination of laminated and homogeneous plate ABD matrix equations, called SMEAR technology, as follows in equations (8.4) through (8.6).

$$[A]_{smear} = \sum_{k=1}^{n} ([\overline{Q}]_k (z_k - z_{k-1}))$$ (8.4)

$$[B]_{smear} = 0$$ (8.5)

$$[D]_{smear} = [A]_{smear} \frac{T^2}{12}$$ (8.6)

It can be shown that the average [D] matrix of all possible combinations of stacking sequences for a given laminate is equal to the [D]smear matrix for that laminate. Therefore, using [D]smear is an appropriate action when you have a laminate with an indeterminate stacking sequence. Problem 8.1 verifies that the average [D] matrix of all possible stacking sequences for a given laminate is indeed equal the [D]smear.

8.3 SMEAR CORE Technology in Composite Design

When designing core-stiffened laminated structures, SMEAR CORE technology should be utilized instead of SMEAR technology. The primary objective of core-stiffened structures is to produce light-weight structures that have high bending stiffness characteristics. In order to efficiently increase the bending stiffness of a structure, material needs to be separated as far as possible from the neutral axis. For traditional metallic structures, this objective is accomplished by I cross-sectional shapes producing the I-beam. However, for laminated plates, this objective is satisfied by separating laminated face-sheets from the neutral axis by means of inserting a light-weight core layer between the laminated face-sheets; thus, making a core-stiffened laminated structure.

The convention for defining core-stiffened laminates is to define ply (1) through ply (n – 1) as all the plies in both laminated face-sheets and to define ply (n) as the core ply. Therefore, the total thickness of both laminated face-sheets is defined by equation (8.7) and the thickness of the core is defined by equation (8.8). Even though this is the convention to define core-stiffened laminates, half of the plies are distributed to the top laminated face-sheet and the other half of the plies are distributed to the bottom laminated face-sheet as shown in figure 8.3.

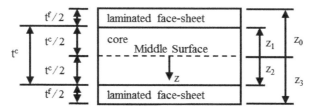

Figure 8.3, Core-Stiffened Laminate Stacking Sequence and Z-Coordinates

$$t^f = \sum_{k=1}^{n-1} t_k \qquad (8.7)$$

$$t^c = t_n \qquad (8.8)$$

The core layer only serves the purpose to separate the laminated face-sheets from the neutral axis and does not add an appreciably stiffness itself. Therefore, the extension behavior of a core-stiffened laminated panel is approximately the same as the extension behavior of the laminated face-sheets themselves.   Using this fact, the $[A]_{smcore}$ matrix for SMEAR CORE technology can be defined as given by equation (8.9). In addition, similar to SMEAR technology, extension-bending coupling is an undesirable behavior; therefore, the $[B]_{smcore}$ matrix for SMEAR CORE technology is set to zero as given by equation (8.10).

$$[A]_{smcore} = \sum_{k=1}^{n-1} ([\overline{Q}]_k (z_k - z_{k-1})) \qquad (8.9)$$

$$[B]_{smcore} = 0 \qquad (8.10)$$

In order to evaluate the $[D]_{smcore}$ matrix for SMEAR CORE technology, we must treat the top and bottom laminated face-sheets as single homogenized laminates as shown in figure 8.3. Since structural bending loads are primarily reacted by internal in-plane tension and compression loads in the laminated face-sheets, it is appropriate to utilize the equivalent in-plane homogenized engineering constant equations, as derived in section 7.2, to calculate the equivalent in-plane homogenized engineering constants of the laminated face-sheets.   Using equation (7.16), the equivalent in-plane homogenized engineering constants of the laminated face-sheets are given by equation (8.11).   It should be noted that the equivalent in-plane homogenized engineering constants of a given laminate are equivalent for any multiple of the given laminate's thickness.   Therefore, the equivalent in-plane homogenized engineering constants of the individual face-sheet laminates are the same as the combined laminate which is double the thickness.

$$E_x^f = \frac{1}{t^f a_{11,smcore}}$$

$$E_y^f = \frac{1}{t^f a_{22,smcore}}$$

$$v_{xy}^f = \frac{-a_{12,smcore}}{a_{11,smcore}} \qquad (8.11)$$

$$v_{yx}^f = \frac{-a_{12,smcore}}{a_{22,smcore}}$$

$$G_{xy}^f = \frac{1}{t^f a_{44,smcore}}$$

Where;

$$[a]_{smcore} = [A]_{smcore}^{-1}$$

In order to calculate the [D] matrix for the core-stiffened laminate, we will consider the core stiffened laminate to be made from three plies; two face-sheet plies and one core ply. Furthermore, calculating the [D] matrix for the core-stiffened laminate using equation (8.3); we need to obtain the plane stress stiffness matrix for the laminated face sheets $[\overline{Q}]^f$, the plane stress stiffness matrix for the core $[\overline{Q}]^c$, and the ply z-coordinate values. Using the in-plane homogenized engineering constants of the laminate face-sheets, the plane stress stiffness matrix of the laminate face-sheets is given by equation (8.12). Since the core layers main purpose is to separate the laminated face-sheets from the neutral axis, it provides no appreciable stiffness itself to the laminate; therefore, $[\overline{Q}]^c = 0$.

$$[\overline{Q}]^f = \begin{bmatrix} \dfrac{E_x^f}{1-v_{xy}^f v_{yx}^f} & \dfrac{v_{xy}^f E_x^f}{1-v_{xy}^f v_{yx}^f} & 0 \\[3mm] \dfrac{v_{xy}^f E_x^f}{1-v_{xy}^f v_{yx}^f} & \dfrac{E_y^f}{1-v_{xy}^f v_{yx}^f} & 0 \\[3mm] 0 & 0 & G_{xy}^f \end{bmatrix} \qquad (8.12)$$

The z-coordinate values for the three plies of the core-stiffened laminate are evaluated using figure 8.3, repeated below for convenience, and are given by equation (8.13).

$$z_0 = \frac{-(t^f + t^c)}{2}$$

$$z_1 = \frac{-t^c}{2}$$ (8.13)

$$z_2 = \frac{t^c}{2}$$

$$z_3 = \frac{(t^f + t^c)}{2}$$

Finally, substituting equations (8.12) and (8.13) into equation (8.3), the [D] matrix for SMEAR CORE technology [D]$_{smcore}$ is given by equation (8.14).

$$[D]_{smcore} = [\overline{Q}]^f \left(\frac{(t^f)(t^c)^2 + (t^f)^2(t^c)}{4} + \frac{(t^f)^3}{12} \right)$$ (8.14)

Chapter 8
Exercises

8.1. Calculate the [A], [B], and [D] matrices for the following laminates. The laminates are made from unidirectional plies of thickness ($t = 0.01$") with transversely isotropic material properties given below.

a) $[0/45/90/-45]_s$
b) $[45/90/-45/0]_s$
c) $[90/-45/0/45]_s$
d) $[-45/0/45/90]_s$

Ply material properties
$E_1 = 22.0e6$ psi
$E_2 = 1.30e6$ psi
$v_{12} = 0.30$
$v_{23} = 0.26$
$G_{12} = 0.75e6$ psi
$G_{23} = 0.516e6$ psi
$\alpha_1 = -0.30e-6$ /°F
$\alpha_2 = 18.0e-6$ /°F

What do you notice about the [A] matrix of the laminates?
What do you notice about the [B] matrix of the laminates?
What do you notice about the [D] matrix of laminates?

8.2. Calculate $[D]_{average}$ for the laminates given in problem 8.1. Calculate $[D]_{smear}$ for the laminates given in problem 8.1. What do you notice about $[D]_{average}$ and $[D]_{smear}$?

8.3. Calculate $[D]_{smcore}$ for the laminates given in problem 8.1. Assume the laminates in problem 8.1 define all the face-sheet plies and that the thickness of the core layer, append to the end of the laminates given in problem 8.1, is $t^c = 0.25$".

8.4. Circle True or False for the following statements. Assume all plies are of the same material and are of equal thickness.

a) The [A] matrix is independent of the stacking sequence.

True   **False**

b) For a balanced laminate, D_{14} and D_{24} terms are always zero.

True   **False**

c) The equivalent in-plane homogenized axial stiffness E_x of a $[90]_{10}$ laminate is greater than the same for a $[90]_5$ laminate.

True   **False**

d) For a symmetric laminate, D_{11} and D_{22} terms will always be the same.

True   **False**

8.5. For the laminate shown below, circle the correct answers.

a) $E_x = E_y$ $E_x < E_y$ $E_x > E_y$

b) $D_{11} = D_{22}$ $D_{11} < D_{22}$ $D_{11} > D_{22}$

c) $A_{14} = 0$ $A_{14} \neq 0$

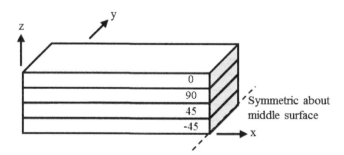

8.6. For the laminate shown below circle the correct answers.

a) $E_x = E_y$ $E_x < E_y$ $E_x > E_y$

b) $D_{11} = D_{22}$ $D_{11} < D_{22}$ $D_{11} > D_{22}$

c) $A_{14} = 0$ $A_{14} \neq 0$

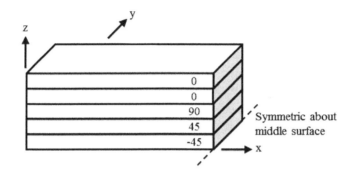

8.7. How would you change the stacking sequence for the laminate shown below to get the maximum D_{44} component value?

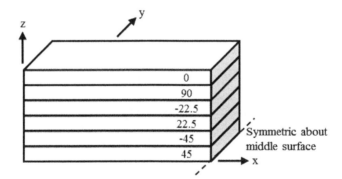

8.8. Consider the laminate shown below. What plies would you add to eliminate the shear deformation which would result from extensional loading? Explain the reasoning behind your answer.

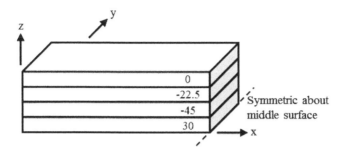

8.9. A laminate twists under a pure bending load M_x. Which stiffness terms from the ABD matrix are causing this twisting behavior? Would you expect this to be more evident in a thick or thin laminate?

8.10. The $0°$ ply has the largest influence on the D_{11} component term. What would the approximate ratio of D_{11} for Laminate A be compared to Laminate B? Assume each ply is a unit thickness. $D_{11}^A / D_{11}^B = ?$

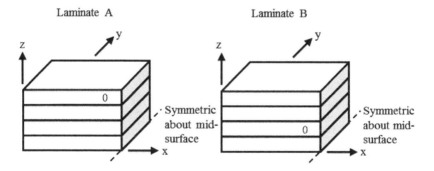

8.11. Put an X in the boxes that apply for each laminate. Assume the same material and thickness for each ply.

Laminate	$[B] = 0$	$A_{14} = A_{24} = 0$	$D_{14} = D_{24} = 0$
[0/90/0/90/0]			
[45/−45/−30/30]			
[0/45/90/45/0]			
[45/−45/90/0/90/−45/45]			
[0/90/−45/45/90/0]			
[90/0/90/0]			
[0/45/−45/0]			
[22.5/−22.5/90/−22.5/22.5]			
[−45/22.5/0/22.5/−45]			
[−30/30/60/30/−30]			

Chapter 9
Composite First Ply Failure Criteria

9.1 Introduction to Structural Failure

There are several challenges facing engineering when it comes to using analytical methods to predict, or even capture the trends, of structural failure. The first challenge is defining; what is failure? The best definition I have come across is by A. Ugural and S. Fenster in their book, *Advanced Strength and Applied Elasticity* (1995). They defined failure with the following statement;

> In the most general terms, failure refers to any action leading to an inability of a part of a structure to function in the manner intended.

Therefore, when discussing failure one must first define; what is failure for their given structure? In this light, typical failure modes for metallic structures include;

- Excessive linear elastic deformation
- Permanent deformation or yielding
- Instability or buckling
- Creep
- Fatigue
- Fracture (the creation of new surfaces)

In a similar manner, typical failure modes for composite structures include;

- Excessive linear elastic deflection
- Instability or buckling
- Creep
- Fatigue
- Matrix crazing and/or shear yielding (onset of failure in the matrix)
- Fiber fracture (onset of failure in the fiber)
- Transverse ply fracture (creation of new surfaces within a ply)
- Ply delamination (creation of new surfaces between plies)

Another challenge facing engineers in predicting structural failure is continuing to increase our knowledge of the physics of material failure mechanisms at the microstructure level and the analytical models we use to model material failure mechanisms at the macroscopic level. In order to discuss our current knowledge of the physics of material failure, it is first necessary to present a few definitions which we will use as standard science and engineering terms throughout this chapter.

Empirical Data

Empirical data is information based on or characterized by observation and experiment.

Empirical Models

Empirical models are working hypothesis, within the bounds of the currently available experimental data that are testable using observation or experiment.

Hypothesis

A tentative explanation for a physical phenomenon used as a basis for further investigation. A hypothesis is assumed to be true until further empirical data either supports (verifies) the hypothesis and elevates it to a theory or contradicts (falsifies) the hypothesis.

Theory

A set of facts, laws, or principles used to explain physical phenomena. Theories are regarded as truth and must be backed by significant empirical evidence supporting (verifying) the theory.

Physics

Physics is the science of matter and its motion. It is the science that seeks to understand the basic concepts of force, energy, mass, and charge. More completely, it is the general analysis of nature, conducted in order to understand how the universe behaves.

A great quote on our current knowledge of the physics of any phenomena is eloquently stated by R. Shankar in his book, *Principles of Quantum Mechanics* (1994);

> Our description of the physical world is dynamic in nature and undergoes frequent change. At any given time, we summarize our knowledge of natural phenomena by means of certain laws (theories and hypothesis). These laws adequately describe the phenomenon studied up to that time, to an accuracy then attainable. As time passes, we enlarge the domain of observation and improve the accuracy of measurement. As we do so, we constantly check to see if the laws continue to be valid. Those laws that do remain valid gain in stature, and those that do not must be abandoned in favor of new ones that do.

Unlike the yielding of ductile metals, for which the physics are well understood by dislocation movement theory at the microscopic level and which is modeled by von Mises yield theory at the macroscopic level, composite material failure mechanisms are not currently well understood at the microscopic level.   Therefore, composite failure models at the macroscopic level have traditionally been empirical curve-fit composite failure models. Empirical curve-fit composite failure models seek to curve-fit an assumed interaction equation given empirical single state failure values. Empirical curve-fit composite failure models have known limitations in their ability to accurately predate material failure mechanisms for composite ply materials.   Still, empirical curve-fit composite failure models are suitable for composite design purposes and are widely used in industry for such applications.

A significant amount of research is currently being focused on expanding our knowledge of the basic composite material failure mechanisms at the microscopic level. This research has led to modern physics based composite failure models at the macroscopic level.   These modern physics based composite failure models are currently being evaluated by both academia and industry; therefore, it is outside the scope of the current edition of this book to discuss these modern physics based composite failure models. However, it should be noted that these modern physics based composite failure models have shown significant improvement in their ability to accurately predict material failure mechanisms over the empirical curve-fit composite failure models. There is a high probability that these physics based composite failure models will be used for more accurate predictive analysis and refined composite design in the future.

9.2 Test Specimens for Stiffness and Strength Properties

The objective of this section is to describe the test specimens that are used to obtain ply level stiffness and strength properties that are necessary for modeling laminate behavior. Unidirectional fiber-matrix ply materials are typically modeled using the linear elastic transversely isotropic material law; therefore, the following engineering stiffness properties are required to be obtained by test;

- E_1 is the Young's modulus in the fiber 1-direction.
- E_2 is the Yong's modulus in the transverse matrix 2-direction.
- v_{12} is the Poisson's ratio in the 12-plane.
- G_{12} is the Shear modulus in the 12-plane.

In addition to the engineering stiffness properties, the engineering strength properties of a unidirectional fiber-matrix ply must also be determined in order to use a first ply failure criterion to predict structural failure by first ply failure. The following engineering strength properties are required to be obtained by test to perform first ply failure analysis;

- X_T is the failure stress/strain in the fiber 1-direction in tension.
- X_C is the failure stress/strain in the fiber 1-direction in compression.
- Y_T is the failure stress/strain in the matrix 2-direction in tension.
- Y_C is the failure stress/strain in the matrix 2-direction in compression.
- S is the failure shear stress/strain in the 12-plane.

All of the engineering stiffness and strength properties are obtained using the following test specimens; 0° tension, 0° compression, 90° tension, 90° compression, and 45/−45 tension. Each test specimen is described below including dimensions and data reduction techniques.

0° Tension Test Specimen

The most common 0° tension test specimen is ASTM D3039 and is used to obtain the following stiffness and strength properties; E_1, v_{12}, and X_T. The typical dimensions for a 0° tension test specimen are given in figure 9.1. The typical thickness for a 0° tension test specimen is in the range of 0.02" - 0.10". The data reduction techniques used for 0° tension test specimens are given below in equations (9.1) through (9.4).

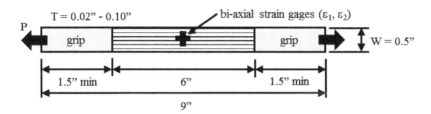

Figure 9.1, 0° Tension Test Specimen

$$\sigma_1 = \frac{P}{WT} \tag{9.1}$$

$$E_1 = \frac{\Delta\sigma_1}{\Delta\varepsilon_1} \text{ in the linear regime of the stress-strain curve.} \tag{9.2}$$

$$v_{12} = \frac{-\Delta\varepsilon_2}{\Delta\varepsilon_1} \text{ in the linear regime of the stress-strain curve.} \tag{9.3}$$

$$X_T = \frac{P_{fail}}{WT} \tag{9.4}$$

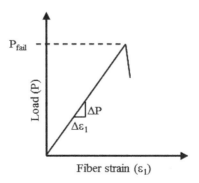

Figure 9.2, Typical 0° Tension Test Load vs. Strain

0° Compression Test Specimen

The most common 0° compression test specimen is ASTM D6641 and is used to obtain the following stiffness and strength properties; E_1 and X_C. The typical dimensions for a 0° compression test specimen are given in figure 9.3. The typical thickness for a 0° compression test specimen is in the range of 0.12" - 0.25". Back-to-back strain gages should be utilized to monitor and verify that excessive bending is not present in the test specimen. Data reduction techniques used for 0° compression test specimens are given below in equations (9.5) through (9.7).

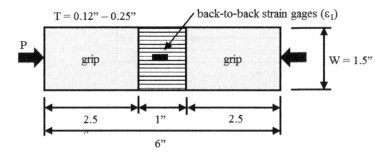

Figure 9.3, 0° Compression Test Specimen

$$\sigma_1 = \frac{P}{WT} \tag{9.5}$$

$$E_1 = \frac{\Delta\sigma_1}{\Delta\varepsilon_1} \text{ in the linear regime of the stress-strain curve.} \tag{9.6}$$

$$X_C = \frac{P_{fail}}{WT} \tag{9.7}$$

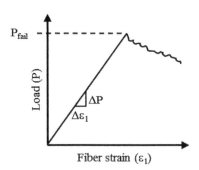

Figure 9.4, Typical 0° Compression Test Load vs. Strain

90° Tension Test Specimen

The most common 90° tension test specimen is ASTM D3039 and is used to obtain the following stiffness and strength properties; E_2, v_{21}, and Y_T. The typical dimensions for a 90° tension test specimen are given in figure 9.5. The typical thickness for a 90° tension test specimen is in the range of 0.02" - 0.10". Data reduction techniques used for 90° tension test specimens are given below in equations (9.8) through (9.11).

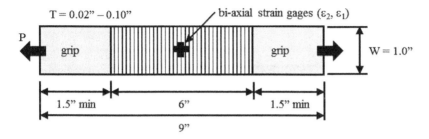

Figure 9.5, 90° Tension Test Specimen

$$\sigma_2 = \frac{P}{WT} \tag{9.8}$$

$$E_2 = \frac{\Delta\sigma_2}{\Delta\varepsilon_2} \text{ in the linear regime of the stress-strain curve.} \tag{9.9}$$

$$v_{21} = \frac{-\Delta\varepsilon_1}{\Delta\varepsilon_2} \text{ in the linear regime of the stress-strain curve.} \tag{9.10}$$

$$Y_T = \frac{P_{fail}}{WT} \tag{9.11}$$

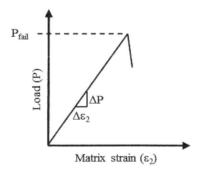

Figure 9.6, Typical 90° Tension Test Load vs. Strain

90° Compression Test Specimen

The most common 90° compression test specimen is ASTM D6641 and is used to obtain the following stiffness and strength properties; E_2 and Y_C. The typical dimensions for a 90° compression test specimen are given in figure 9.7. The typical thickness for a 90° compression test specimen is in the range of 0.12" - 0.25". Back-to-back strain gages should be utilized to monitor and verify that excessive bending is not present in the test specimen. Data reduction techniques used for 90° compression test specimens are given below in equations (9.12) through (9.14).

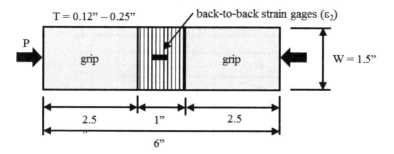

Figure 9.7, 90° Compression Test Specimen

$$\sigma_2 = \frac{P}{WT} \tag{9.12}$$

$$E_2 = \frac{\Delta\sigma_2}{\Delta\varepsilon_2} \text{ in the linear regime of the stress-strain curve.} \tag{9.13}$$

$$Y_C = \frac{P_{fail}}{WT} \tag{9.14}$$

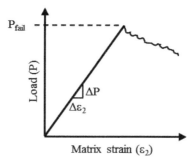

Figure 9.8, Typical 90° Compression Test Load vs. Strain

+/−45 Tension Test Specimen

The most common +/−45 tension specimen is ASTM D3518 and is used to obtain the following shear stiffness and shear strength properties; G_{12} and S. The typical dimensions for a +/−45 tension test specimen are given in figure 9.9. The typical stacking sequence for a +/−45 tension test specimen is $[(45/−45)_n]_s$ with typical thickness in the range of 0.02" - 0.10". Data reduction techniques used for +/−45 tension test specimens are given below in equations (9.15) through (9.18).

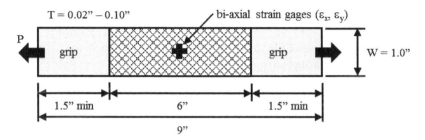

Figure 9.9, 45/−45 Tension Specimen

$$\tau_{12} = \frac{P}{2WT} \tag{9.15}$$

$$\gamma_{12} = \varepsilon_x - \varepsilon_y \tag{9.16}$$

$$G_{12} = \frac{\Delta\tau_{12}}{\Delta\gamma_{12}} \text{ in the linear regime of the shear stress-strain curve.} \tag{9.17}$$

$$S = \frac{P_{fail}}{2WT} \tag{9.18}$$

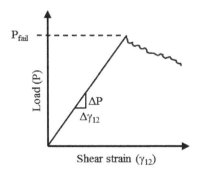

Figure 9.10, Typical 45/−45 Tension Text Load vs. Strain

9.3 Empirical Curve-Fit First Ply Failure Criteria

This section reviews empirical curve-fit first ply failure criteria that are traditionally utilized and implemented in most commercial finite element solvers. The objective of empirical curve-fit criteria is to obtain a curve-fit to an assumed interaction equation given empirical single state failure values, as demonstrated in by the Tsai-Hill first ply failure criteria. Empirical curve-fit first ply failure criteria have known limitations in their ability to accurately predate the first ply failure mechanism for composite ply materials, due to the assumed interaction equation inaccurate representation of the physics of composite first ply failure behavior. Still, empirical curve-fit first ply failure criteria are suitable for composite design purposes and are widely used in industry for such applications. The empirical curve-fit first ply failure criteria discussed in this section include; maximum stress, maximum strain, Tsai-Hill, and Tsai-Wu first ply failure criterion.

Maximum Stress First Ply Failure Criteria

The maximum stress first ply failure criteria can be written as given by equation (9.19). The maximum stress first ply failure envelope is graphically represented in figure 9.6, for a shear stress $\tau_{12} = 0.0$, and provides the largest failure envelope of all the empirical curve-fit first ply failure criteria. The maximum stress first ply failure criteria has known limitations capturing the first ply failure event in combined stress states.

$$\max\left(\left|\frac{\sigma_1}{X}\right|, \left|\frac{\sigma_2}{Y}\right|, \left|\frac{\tau_{12}}{S}\right|\right) = 1 \qquad (9.19)$$

Where;
$X = X_T$ if $\sigma_1 > 0$; else $X = X_C$ if $\sigma_1 < 0$.
$Y = Y_T$ if $\sigma_2 > 0$; else $Y = Y_C$ if $\sigma_2 < 0$.

Maximum Strain First Ply Failure Criteria

The maximum strain first ply failure criteria can be written as given by equation (9.20). The maximum strain first ply failure envelope is similar to the maximum stress first ply failure envelope which is graphically represented in figure 9.6, for a shear stress $\tau_{12} = 0.0$, and provides the largest failure envelope of all the empirical curve-fit first ply failure criteria. The maximum strain first ply failure criteria has known limitations capturing the first ply failure event in combined stress states.

$$\max\left(\left|\frac{\varepsilon_1}{X}\right|, \left|\frac{\varepsilon_2}{Y}\right|, \left|\frac{\gamma_{12}}{S}\right|\right) = 1 \tag{9.20}$$

Where;
$X = X_T$ if $\varepsilon_1 > 0$; else $X = X_C$ if $\varepsilon_1 < 0$.
$Y = Y_T$ if $\varepsilon_2 > 0$; else $Y = Y_C$ if $\varepsilon_2 < 0$.

Tsai-Hill First Ply Failure Criteria

The Tsai-Hill first ply failure criteria can be written in its most generic form as given by equation (9.21).

$$(G + H)\sigma_1^2 + (F + H)\sigma_2^2 + (F + G)\sigma_3^2 - 2H\sigma_1\sigma_2 - 2G\sigma_1\sigma_3$$
$$- 2F\sigma_2\sigma_3 + 2L\tau_{23}^2 + 2M\tau_{13}^2 + 2N\tau_{12}^2 = 1 \tag{9.21}$$

In order to determine the constants E, F, G, H, L, M, and N; six boundary conditions must be prescribed as follows.

Boundary condition 1
$$\tau_{12} = S_{12} \tag{9.22}$$
$$\sigma_1 = \sigma_2 = \sigma_3 = \tau_{23} = \tau_{13} = 0$$

Boundary condition 2
$$\tau_{23} = S_{23} \tag{9.23}$$
$$\sigma_1 = \sigma_2 = \sigma_3 = \tau_{12} = \tau_{13} = 0$$

Boundary condition 3
$$\tau_{13} = S_{13} \tag{9.24}$$
$$\sigma_1 = \sigma_2 = \sigma_3 = \tau_{12} = \tau_{23} = 0$$

Boundary condition 4
$$\sigma_1 = X \tag{9.25}$$
$$\sigma_2 = \sigma_3 = \tau_{23} = \tau_{13} = \tau_{12} = 0$$

Boundary condition 5
$$\sigma_2 = Y \tag{9.26}$$
$$\sigma_1 = \sigma_3 = \tau_{23} = \tau_{13} = \tau_{12} = 0$$

Boundary condition 6
$$\sigma_3 = Z \tag{9.27}$$
$$\sigma_1 = \sigma_2 = \tau_{23} = \tau_{13} = \tau_{12} = 0$$

Substituting boundary conditions 1, 2, and 3 into equation (9.21) yields the constants L, M, and N as given by equations (9.28) through (9.30) respectively.

$$2N = \frac{1}{S^2} \tag{9.28}$$

$$2L = \frac{1}{S^2} \tag{9.29}$$

$$2M = \frac{1}{S^2} \tag{9.30}$$

Substituting boundary conditions 4, 5, and 6 into equation (9.21) yields the system of equations (9.31) through (9.33) involving the constants F, G, and H.

$$(G + H) = \frac{1}{X^2} \tag{9.31}$$

$$(F + H) = \frac{1}{Y^2} \tag{9.32}$$

$$(F + G) = \frac{1}{Z^2} \tag{9.33}$$

Solving the system of equations (9.31) through (9.33) yields the constants F, G, and H as equations (9.34) through (9.36) respectively.

$$2F = \frac{-1}{X^2} + \frac{1}{Y^2} + \frac{1}{Z^2} \tag{9.34}$$

$$2G = \frac{1}{X^2} - \frac{1}{Y^2} + \frac{1}{Z^2} \tag{9.35}$$

$$2H = \frac{1}{X^2} + \frac{1}{Y^2} - \frac{1}{Z^2} \tag{9.36}$$

Substituting the constants F, G, H, L, M, and N back into equation (9.21), the Tsai-Hill first ply failure criteria in 3D space is derived as given by equation (9.37).

$$\frac{\sigma_1^2}{X^2} + \frac{\sigma_2^2}{Y^2} + \frac{\sigma_3^2}{Z^2} - \left(\frac{1}{X^2} + \frac{1}{Y^2} - \frac{1}{Z^2}\right)\sigma_1\sigma_2 - \left(\frac{1}{X^2} - \frac{1}{Y^2} + \frac{1}{Z^2}\right)\sigma_1\sigma_3 \tag{9.37}$$
$$- \left(\frac{-1}{X^2} + \frac{1}{Y^2} + \frac{1}{Z^2}\right)\sigma_2\sigma_3 + \frac{\tau_{23}^2}{S_{23}^2} + \frac{\tau_{13}^2}{S_{13}^2} + \frac{\tau_{12}^2}{S_{12}^2} = 1$$

Assuming a plane stress environment, $\sigma_3 = \tau_{23} = \tau_{13} = 0$, and equal strengths in the matrix 2-direction and through-thickness 3-direction; the Tsai-Hill first ply failure criteria in plane stress for a unidirectional fiber-matrix ply material is derived as given by equation (9.38).

$$\frac{\sigma_1^2}{X^2} + \frac{\sigma_2^2}{Y^2} - \frac{\sigma_1\sigma_2}{X^2} + \frac{\tau_{12}^2}{S^2} = 1 \tag{9.38}$$

Where;
$X = X_T$ if $\sigma_1 > 0$; else $X = X_C$ if $\sigma_1 < 0$.
$Y = Y_T$ if $\sigma_2 > 0$; else $Y = Y_C$ if $\sigma_2 < 0$.

The Tsai-Hill first ply failure envelope is graphically represented in figure 9.6, for a shear stress $\tau_{12} = 0.0$, and provides a more accurate representation of the first ply failure event. Still, the Tsai-Hill first ply failure criteria has known limitations capturing the first ply failure event in combined stress states. Also, the Tsai-Hill first ply failure criteria does not accurately simulate materials that exhibit differences in tension and compression properties. For this reason, the Tsai-Wu first ply failure criteria is typically utilized over the Tsai-Hill first ply failure criteria. However, under certain circumstances as will be discussed below, the Tsai-Wu first ply failure criteria reduces to the Tsai-Hill first ply failure criteria.

Tsai-Wu First Ply Failure Criteria

The Tsai-Wu first ply failure criteria can be written in its most generic form as given by equation (9.39).

$$F_{ij}\sigma_i\sigma_j + F_i\sigma_i = 1 \tag{9.39}$$

Expanding equation (9.39) for a plane stress environment, $\sigma_3 = \tau_{23} = \tau_{13} = 0$, the Tsai-Wu first ply failure criteria for plane stress is given as equation (9.40).

$$F_{11}\sigma_1^2 + 2F_{12}\sigma_1\sigma_2 + F_{22}\sigma_2^2 + F_{44}\tau_{12}^2 + 2F_{14}\sigma_1\tau_{12} + 2F_{24}\sigma_2\tau_{12} \tag{9.40}$$
$$+ F_1\sigma_1 + F_2\sigma_2 + F_4\tau_{12} = 1$$

Considering that the strength should be unaffected by the sign of the shear stress component, all linear terms in τ_{12} must vanish; therefore, $F_{14} = F_{24} = F_4 = 0$ leaving equation (9.41).

$$F_{11}\sigma_1^2 + 2F_{12}\sigma_1\sigma_2 + F_{22}\sigma_2^2 + F_{44}\tau_{12}^2 + F_1\sigma_1 + F_2\sigma_2 = 1 \tag{9.41}$$

Of the six remaining material constant terms, F_{11}, F_{22}, F_{44}, F_1, and F_2 can readily be determined by substituting the boundary conditions 1 through 6, as given above for the Tsai-Hill first ply failure criteria, into equation (9.41). The results of these substitutions and solving for the unknown material constants are given below as equations (9.42) through (9.46). The material constant term F_{12} cannot be readily determined in a simply analytical way, and must be determined by a bi-axial stress test specimen for which no standard currently exists.

$$F_{11} = \frac{1}{X_T X_C} \tag{9.42}$$

$$F_{22} = \frac{1}{Y_T Y_C} \tag{9.43}$$

$$F_{44} = \frac{1}{S^2} \tag{9.44}$$

$$F_1 = \frac{1}{X_T} - \frac{1}{X_C} \tag{9.45}$$

$$F_2 = \frac{1}{Y_T} - \frac{1}{Y_C} \tag{9.46}$$

Substituting the material constant terms back into equation (9.41) one obtains the Tsai-Wu first ply failure criteria for plane stress as given by equation (9.47).

$$\frac{\sigma_1^2}{X_T X_C} + 2F_{12}\sigma_1\sigma_2 + \frac{\sigma_2^2}{Y_T Y_C} + \frac{\tau_{12}^2}{S^2} + \left(\frac{1}{X_T} - \frac{1}{X_C}\right)\sigma_1 + \left(\frac{1}{Y_T} - \frac{1}{Y_C}\right)\sigma_2 = 1$$

$$(9.47)$$

The Tsai-Wu first ply failure envelope is graphically represented in figure 9.11, for a shear stress $\tau_{12} = 0.0$, and provides the most accurate representation of the first ply failure event of all empirical curve-fit first ply failure criteria. Still, the Tsai-Wu first ply failure criterion has known limitations capturing the first ply failure event in combined stress states. A couple things to note about the Tsai-Wu first ply failure criteria.

1. Tsai-Wu accounts for differences between tension and compression failure strengths.
2. In general, F_{12} cannot be analytically determined, and it is common to set $F_{12} = 0$ in practice.
3. The Tsai-Wu first ply failure criteria equals the Tsai-Hill first ply failure criteria if the tension and compression strengths are equal and $F_{12} = -1/2X^2$.

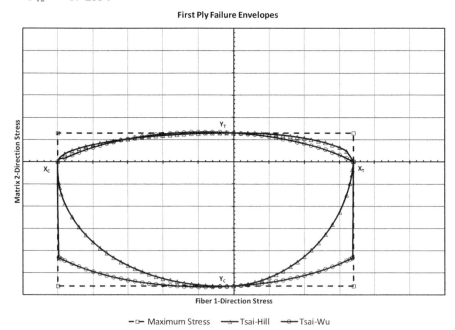

First Ply Failure Envelopes

Maximum Stress ---□---   Tsai-Hill ---▲---   Tsai-Wu ---●---

Figure 9.11, Empirical Curve-Fit First Ply Failure Envelopes

Chapter 9
Exercises

9.1. Using the unidirectional ply material strengths given below, calculate the Tsai-Wu first ply failure index for the 90° ply of the laminate in problem 6.1 under the boundary condition $N_x = 2400 lbs/in$ as in problem 6.4.

Ply material strengths
$X_T = 170,000$
$X_C = 170,000$
$Y_T = 6,500$
$Y_C = 28,000$
$S = 10,000$

9.2. Using the unidirectional ply material strengths as given in problem 9.1, calculate the Tsai-Wu first ply failure index for the 90° ply of the laminate in problem 6.1 under the boundary condition $\Delta T = -100°F$ as in problem 6.6.

9.3. Using the unidirectional ply material strengths as given in problem 9.1, calculate the Tsai-Wu first ply failure index for the 90° ply of the laminate in problem 6.1 under the combined boundary condition $N_x = 2400 lbs/in$ and $\Delta T = -100°F$ as in problem 6.7.

Chapter 10
Composite Analysis

This chapter reviews the general composite model building process for current common finite element solvers. An example laminated plate model, defined in figure 10.1, will be used to demonstrate the three general composite model building processes using the finite element method; zone-based shell modeling, ply-based shell modeling, and ply-by-ply solid modeling. The example laminated plate model shown in figure 10.1 has both tria and quad shell elements (pyramid and hex solid elements). Each element has a material coordinate system, which defines the 0° fiber direction for each element. In addition, figure 10.1 defines the boundary conditions, laminate stacking sequence, and transversely isotropic engineering constants of the ply material from which the laminate is made. The boundary conditions include both mechanical and thermal loadings so as to generalize the problem and allow for comparison with the classical lamination plate theory analysis of chapter 6.

Composite shell zone-based modeling is the traditional approach to building composite models. It requires a property definition for each laminate zone within a composite structure. Therefore, at each ply drop or addition location, another laminate zone property definition must be defined. Each laminate zone property definition must completely define the laminate within that laminate zone. Plies that extend through multiple laminate zones must be redefined in each laminate zone property definition. The duplication of ply data within each laminate zone property definition causes inefficient data handling and ultimately necessitated the need for developing an alternative composite modeling approach. This lead to the composite shell ply-based modeling approach discussed next.

Composite shell ply-based modeling is a relatively new technique for building composite models that attempts to mimic the composite manufacturing process of cutting and stacking plies "on top" of each other to construct a composite structure. In the composite ply-based modeling approach, plies are defined as physical entities of a given material with thickness and shape. Plies are then stacked "on top" of each other in a given specified order. In this way, a ply is defined only once. In addition, since the ply shape for all plies are known, the laminate zones are automatically derived. The principal advantage of ply-based modeling is the ability to easily make design updates to composite models via addition and subtraction of plies and modification of ply shapes, which automatically recalculates the laminate zones for the composite model.

Composite ply-by-ply solid modeling is the most accurate modeling technique and is the only method which can accurately capture through-thickness effects. However, ply-by-ply solid modeling requires a solid layer of elements for each ply. Given that plies are relatively thin and that finite element quality metrics limit h-element aspect ratios to be no greater than 1:5, a very large number of h-elements are typically required to accurately model composite structures with ply-by-ply solid methods. Regardless of the complexity associated with developing these models and since it is the only method to accurately capture through-thickness effects, ply-by-ply solid models are still commonly developed for the design and analysis of composite structural joints.

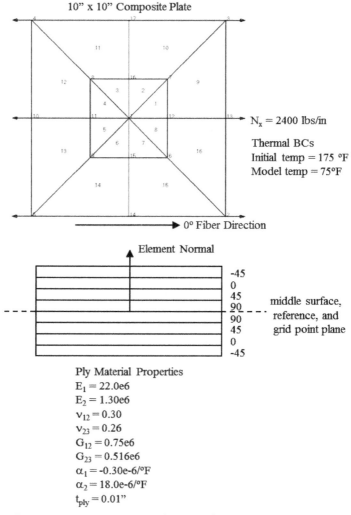

Figure 10.1, Composite Plate Analysis Model

10.1 Composite Zone-Based Shell Modeling

The general process to develop a composite zone-based shell model for the design and analysis of composite structures is described in this section. The modeling procedures discussed are generally generic to all finite element solvers, however we will be explicitly developing models using the finite element solver OptiStruct. The required OptiStruct i/o options, subcase control, and bulk data necessary to build a composite zone-based shell model can be referenced in Appendices A, B, and C respectively. The final OptiStruct composite zone-based shell model, as given in figure 10.1, can be referenced in Appendix E – Composite Zone-Based Shell Model.

1. **Create shell elements** by generating a finite element mesh using a suitable finite element pre-processor (Altair HyperMesh). Composite shell elements are defined in OptiStruct through the GRID, CTRIA3, and CQUAD4 bulk data cards.

2. **Assign element normal.** An element normal defines the laminate stacking sequence direction for a given element. The laminate stacking sequence is given as the order in which the plies are defined within a laminate zone property definition. Typically ply 1 is the first ply defined and ply n is the last ply defined within a laminate zone property definition. Plies stack in the direction of the element normal from ply 1 to ply n. Therefore, defining the element normal correctly is of critical importance. It is recommended that users consult the specific solver documentation for the element normal conventions used within that solver. An element normal is defined in OptiStruct by the order of the grids on the CTRIA3 and CQUAD4 bulk data cards.

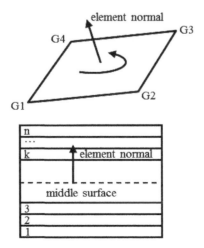

Figure 10.2, Zone-Based Element Normal and Stacking Sequence Definitions

3. **Assign element material coordinate system.** The element material coordinate system x-axis defines the direction of a 0° ply for an element. Furthermore, the ply fiber direction θ_k, defined for each kth ply within a laminate zone property definition, is always relative to the element material coordinate system x-axis as shown in figure 10.3. There are several methods for defining the element material coordinate system for various finite element analysis solvers. However, most solvers define the element material coordinate system x-axis as an angle θ from the G1-G2 vector about the element normal as shown in figure 10.3. It is recommended that users consult the specific solver documentation for the element material coordinate system conventions used within that solver. An element material coordinate system is defined in OptiStruct through the θ field on the CTRIA3 or CQUAD4 bulk data cards.

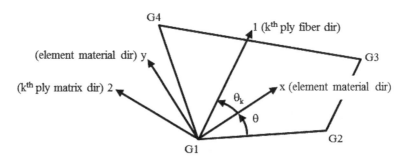

Figure 10.3, Zone-Based Element Material System and Ply Orientation

4. **Create homogeneous ply materials** for each unique ply material that is utilized within the laminates that make up the composite structure. In general, most solvers support the creation of plane stress isotropic, transversely isotropic, and orthotropic homogeneous ply materials as per the definitions in chapter 3. It is recommended that users consult the specific solver documentation for the definitions of the ply materials that can be created within that solver. Homogeneous ply materials are defined in OptiStruct through the MAT1 (isotropic), MAT2 (anisotropic), or MAT8 (orthotropic) bulk data cards.

5. **Create laminate zone property definition** for each laminate zone of the composite structure. A laminate zone is a constant thickness zone of the laminate. Laminate zone boundaries are defined by ply shape boundaries. At each ply shape boundary a new laminate constant thickness zone must exist. Laminate zones are defined as properties in most finite element analysis solvers. The laminate stacking sequence within a laminate zone is given as the order in which the plies are defined within the laminate

zone property definition. Typically ply 1 is the first ply defined and ply n is the last ply defined within a laminate zone property definition. Plies stack in the direction of the element normal from ply 1 to ply n. See figure 10.2 for a schematic of a laminate zone property stacking sequence definition. For each ply defined in a laminate zone property definition, the following ply data is typically required;

- Ply global identification number, GPLYID filed on the PCOMPG bulk data card defines a unique id for each ply.
- Ply material identification number, MID field on the PCOMPG bulk data card defines the plane stress stiffness matrix in the material coordinate system [Q] for the ply.
- Ply thickness, t_k field on the PCOMPG bulk data card.
- Ply fiber direction, θ_k field on the PCOMPG bulk data card. The ply fiber direction is always relative to the element material coordinate system x-axis as defined in figure 10.3.
- Ply results, SOUT field on the PCOMPG bulk data card determines whether or not to calculate and output results for the ply.

It is recommended that users consult the specific solver documentation for the laminate zone property definition conventions used within that solver. Laminate zone properties are defined in OptiStruct through the PCOMP and PCOMPG bulk data cards. It is generally recommend to use the PCOMPG bulk data card over the PCOMP bulk data card.

6. **Assign laminate zone properties to the elements** that represent the laminate zones defined by the laminate zone property definitions. This process assigns an element its stiffness matrix, in this case the ABD matrix of the laminate zone that the element is within. Laminate zone properties are assigned to elements in OptiStruct through the PID field on the CTRIA3 or CQUAD4 bulk data cards.

7. **Create boundary conditions** applied to the composite model that simulate the in-situ environments under investigation.

- Constraints are defined in OptiStruct through the SPC bulk data card.
- Forces and Moments are defined in OptiStruct through the FORCE and MOMENT bulk data cards respectively.
- Pressures on shell elements are defined in OptiStruct through the PLOAD2 bulk data card.
- Initial and Model temperature distributions are defined in OptiStruct through the TEMP or TEMPD bulk data cards.

8. **Create load steps** for each load case that the composite model is to be analyzed for by combining appropriate boundary conditions that simulate the in-situ environments of the composite structure under investigation. Load steps are defined in OptiStruct through the SUBCASE, ANALYSIS, TITLE, SPC, LOAD, and TEMPERATURE(LOAD) subcase control cards.

9. **Create control cards** to define initial temperatures, output results, output formats, and solver controls.

   - The initial temperature distribution is defined in OptiStruct through the TEMPERATURE(INITIAL) subcase control card.
   - Displacement output is defined in OptiStruct through the DISPLACEMENT i/o options card.
   - Composite ply-level strain output is defined in OptiStruct through the CSTRAIN i/o options card. If thermal boundary conditions are applied, then it is important to request ply-level mechanical strain output using the MECH option on the CSTRAIN i/o options card.
   - Composite ply-level stress output is defined in OptiStruct through the CSTRESS i/o options card.
   - Composite ply-level failure index output is defined in OptiStruct through the SB and FT fields on the PCOMP or PCOMPG bulk data cards.
   - Output file formats are defined in OptiStruct through the OUPTUT i/o options card. Typically H3D (binary file for post-processing in Altair HyperView) and ASCII (text file) file formats are requested.

10. **Export the solver input file** representing the composite analysis model from the pre-processor (HyperMesh) **and solve** the composite analysis by submitting the solver input file to the solver executable (OptiStruct).

11. **Post-process the composite analysis results**. The most important results for composite models are the ply-level mechanical strains and stresses in the material coordinate system. It is recommended that users consult the specific solver documentation to determine the exact coordinate system and type of strain (total, mechanical, and/or thermal) that are output from that solver. Most solvers, by default, output ply-level results in the material coordinate system and total strain tensors. Note that the mechanical strain tensors must be used whenever there is a thermal boundary condition applied. If there is no thermal boundary condition applied, then the mechanical strain tensor is equivalent to the total strain tensor and the default output from most solvers can be used directly.

10.2 Composite Ply-Based Shell Modeling

The general process to develop a composite ply-based shell model for the design and analysis of composite structures is described in this section. The modeling procedures discussed are generic to all finite element solvers, however we will be explicitly developing models using the finite element solver OptiStruct. The required OptiStruct i/o options, subcase control, and bulk data necessary to build a composite ply-based shell model can be referenced in Appendices A, B, and C respectively. The final OptiStruct composite ply-based shell model, as given in figure 10.1, can be referenced in Appendix E – Composite Ply-Based Shell Model.

1. **Create shell elements** by generating a finite element mesh using a suitable finite element pre-processor (Altair HyperMesh). Composite shell elements are defined in OptiStruct through the GRID, CTRIA3, and CQUAD4 bulk data cards.

2. **Assign element normal.** An element normal defines the laminate stacking sequence direction for a given element. The laminate stacking sequence is given as the order in which the plies are defined within a laminate zone property definition. Typically ply 1 is the first ply defined and ply n is the last ply defined within a laminate zone property definition. Plies stack in the direction of the element normal from ply 1 to ply n. Therefore, defining the element normal correctly is of critical importance. It is recommended that users consult the specific solver documentation for the element normal conventions used within that solver. An element normal is defined in OptiStruct by the order of the grids on the CTRIA3 and CQUAD4 bulk data cards.

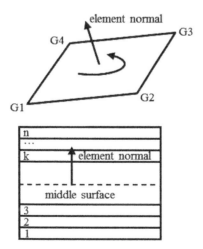

Figure 10.4, Ply-Based Element Normal and Stacking Sequence Definitions

3. **Assign element material coordinate system.** The element material coordinate system x-axis defines the direction of a 0° ply for an element. Furthermore, the ply nominal fiber direction θ_k, defined for each kth ply, is always relative to the element material coordinate system x-axis as shown in figure 10.5. Also, the ply actual fiber direction θ_i, defined for each element of the ply shape, is always relative to the nominal fiber direction and defines the actual fiber direction for the ith element of the kth ply. There are several methods for defining the element material coordinate system for various finite element analysis solvers. However, most solvers define the element material coordinate system x-axis as an angle θ from the G1-G2 vector about the element normal as shown in figure 10.5. It is recommended that users consult the specific solver documentation for the element material coordinate system conventions used within that solver. An element material coordinate system is defined in OptiStruct through the θ field on the CTRIA3 or CQUAD4 bulk data cards.

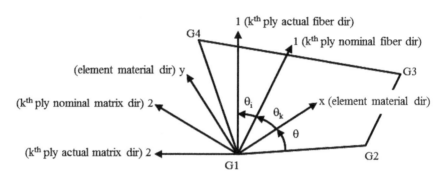

Figure 10.5, Ply-Based Element Material System and Ply Orientation

4. **Create homogeneous ply materials** for each unique ply material that is utilized within the laminates that make up the composite structure. In general, most solvers support the creation of plane stress isotropic, transversely isotropic, and orthotropic homogeneous ply materials as per the definitions in chapter 3. It is recommended that users consult the specific solver documentation for the definitions of the ply materials that can be created within that solver. Homogeneous ply materials are defined in OptiStruct through the MAT1 (isotropic), MAT2 (anisotropic), or MAT8 (orthotropic) bulk data cards.

5. **Create plies** that make up the composite structure. The principal difference between a ply and a ply definition within a laminate zone property definition of zone-based shell modeling is that a ply additionally defines the ply shape along with the ply data of material, thickness, and

fiber direction. It is the ply shape data that is the critical piece of data that allows for the automatic calculation of the composite laminate zones. By defining a ply in this way, building composite models is exactly analogous to the composite structures manufacturing process where a ply is cut to shape and then stacked to build up a laminated composite structure. For each ply the following ply data is typically required;

- Ply material identification number, MID field on the PLY bulk data card defines the plane stress stiffness matrix in the material coordinate system [Q] for the ply.
- Ply thickness, t_k field on the PLY bulk data card.
- Ply nominal fiber direction, θ_k field on the PLY bulk data card. The ply fiber direction is always relative to the element material coordinate system x-axis as defined in figure 10.5.
- Ply actual fiber direction, θ_i field defined on the DRAPE bulk data card for each element of the ply shape. The DID field on the PLY bulk data card references the drape table which defines the actual fiber directions for each element of the ply. The ply actual fiber direction is always relative to the ply nominal fiber direction as defined in figure 10.5. Typically the ply actual fiber direction θ_i is used to interface with draping solvers and obtain more accurate fiber directions for the ply as it is actually manufactured on the final part.
- Ply results, SOUT field on the PLY bulk data card determines whether or not to calculate and output results for the ply.
- Ply shape. The ply shape is typically defined by a set of elements that represent the actual ply shape on the mesh of the composite structure and is defined by the ESID field on the PLY bulk data card.

It is recommended that users consult the specific solver documentation for the ply definitions used within that solver. Plies are defined in OptiStruct through the PLY bulk data card. The actual fiber orientation angle for a ply is defined in OptiStruct through the DRAPE bulk data card.

6. **Create laminates** by stacking plies in a given stacking sequence order. Typically ply 1 is the first ply defined and ply n is the last ply defined within a stack definition. Plies stack in the direction of the element normal from ply 1 to ply n as shown in figure 10.4. It is recommended that users consult the specific solver documentation for the stacking-sequence conventions used within that solver. Laminates are defined in OptiStruct through the STACK bulk data card.

7. **Create ply-based properties**. In zone-based composite modeling, a laminated zone property definition defines a laminate zone. On assignment of a laminated zone property definition to an element; the element stiffness matrix is completely defined. However, in the case of ply-based modeling, a ply-based property is simply a template property defining element level laminate property definitions, such as element offset defined by Z0 on the PCOMPP bulk data card. Assignment of a ply-based property to an element "tags" the element as having a ply-based laminate definition. Each elements actual property is then automatically resolved from the ply and stack definitions that are defined by the PLY and STACK bulk data cards above. Ply-based properties are defined in OptiStruct through the PCOMPP bulk data card.

8. **Assign ply-based properties to the elements** that represent the ply-based composite structure. This process "tags" the element as having a ply-based laminate definition. Each elements actual property is then automatically resolved from the ply and stack definitions that are defined by the PLY and STACK bulk data cards above into the elements stiffness matrix, the ABD matrix in the case of composite elements. Ply-based properties are assigned to elements in OptiStruct through the PID field on the CTRIA3 or CQUAD4 bulk data cards.

9. **Create boundary conditions** applied to the composite model that simulate the in-situ environments under investigation.

   - Constraints are defined in OptiStruct through the SPC bulk data card.
   - Forces and Moments are defined in OptiStruct through the FORCE and MOMENT bulk data cards respectively.
   - Pressures on shell elements are defined in OptiStruct through the PLOAD2 bulk data card.
   - Initial and Model temperature distributions are defined in OptiStruct through the TEMP or TEMPD bulk data cards.

10. **Create load steps** for each load case that the composite model is to be analyzed for by combining appropriate boundary conditions that simulate the in-situ environments of the composite structure under investigation. Load steps are defined in OptiStruct through the SUBCASE, ANALYSIS, TITLE, SPC, LOAD, and TEMPERATURE(LOAD) subcase control cards.

11. **Create control cards** to define initial temperatures, output results, output formats, and solver controls.

- The initial temperature distribution is defined in OptiStruct through the TEMPERATURE(INITIAL) subcase control card.
- Displacement output is defined in OptiStruct through the DISPLACEMENT i/o options card.
- Composite ply-level strain output is defined in OptiStruct through the CSTRAIN i/o options card. If thermal boundary conditions are applied, then it is important to request ply-level mechanical strain output using the MECH option on the CSTRAIN i/o options card.
- Composite ply-level stress output is defined in OptiStruct through the CSTRESS i/o options card.
- Composite ply-level failure index output is defined in OptiStruct through the SB and FT fields on the PCOMP or PCOMPG bulk data cards.
- Output file formats are defined in OptiStruct through the OUPTUT i/o options card. Typically H3D (binary file for post-processing in Altair HyperView) and ASCII (text file) file formats are requested.

12. **Export the solver input file** representing the composite analysis model from the pre-processor (HyperMesh) and solve the composite analysis by submitting the solver input file to the solver executable (OptiStruct).

13. **Post-process the composite analysis results**. The most important results for composite models are the ply-level mechanical strains and stresses in the material coordinate system. It is recommended that users consult the specific solver documentation to determine the exact coordinate system and type of strain (total, mechanical, and/or thermal) that are output from that solver. Most solvers, by default, output ply-level results in the material coordinate system and total strain tensors. Note that the mechanical strain tensors must be used whenever there is a thermal boundary condition applied. If there is no thermal boundary condition applied, then the mechanical strain tensor is equivalent to the total strain tensor and the default output from most solvers can be used directly.

At first glance, it may appear that the composite ply-based modeling method is more cumbersome than the composite zone-based modeling method based solely on the number of steps required to build the composite models between the two methods. However, upon modification of any composites model, due to a design update, the efficiency of the composite ply-based modeling techniques become readily apparent. This will be demonstrated by the exercise problems at the end of this chapter. In addition, composite ply-based modeling techniques have significant advantages for composite design optimization which is the topic of Chapter 11.

10.3 Composite Ply-by-Ply Solid Modeling

The general process to develop a composite ply-by-ply solid model for the design and analysis of composite structures is described in this section. The modeling procedures discussed are generic to all finite element solvers, however we will be explicitly developing models using the finite element solver OptiStruct. The required OptiStruct i/o options, subcase control, and bulk data necessary to build a composite ply-by-ply solid model can be referenced in Appendices A, B, and C respectively. The final OptiStruct composite ply-by-ply solid model, as given in figure 10.1, can be referenced in Appendix E – Composite Ply-by-Ply Solid Model.

1. **Create solid elements** by generating a finite element mesh for each ply using a suitable finite element pre-processor (Altair HyperMesh). A layer of solid elements is required for each ply. Given that plies are typically 0.01" thick and that finite element quality metrics limit h-element aspect ratios to be no greater than 1:5, an h-element length of no greater than 0.05" is typically required. This requirement typically produces a very large number of h-elements to accurately model composite structures with ply-by-ply solid methods. Finite Element p-element formulation such as ESRD StressCheck, overcome this limitation and could be a more suitable choice for composite ply-by-ply solid modeling of complex parts. Composite solid elements are defined in OptiStruct through the GRID, CPENTA, and CHEXA bulk data cards.

2. **Create a material coordinate system** for each unique ply fiber direction. The ply fiber 1-direction is always the material coordinate system x-axis for a rectangular coordinate system. The ply matrix 2-direction is always the material coordinate system y-axis for a rectangular coordinate system. Finally, the ply through-thickness 3-direction is always the material coordinate system z-axis for a rectangular coordinate system. Material coordinate systems are defined in OptiStruct with the COORD1R or CORD2R bulk data cards.

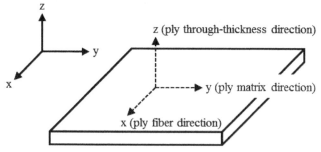

Figure 10.6, Solid Element Material Coordinate System Definition

3. **Create a homogeneous solid ply material** for each unique ply material that is utilized within the laminates that make up the composite structure. In general, most solvers support the creation of isotropic, transversely isotropic, and orthotropic homogeneous ply materials as per the definitions in chapter 3. It is recommended that users consult the specific solver documentation for the definitions of the ply materials that can be created within that solver. Homogeneous ply materials are defined in OptiStruct through the MAT1 (isotropic), MAT9ORT (orthotropic), or MAT9 (anisotropic) bulk data cards.

4. **Create a solid ply property definition** for each ply of the composite structure. Ply properties are defined in OptiStruct through the PSOLID bulk data card. For each ply the following solid ply property data is typically required;

   - Ply material identification number, MID field on the PSOLID bulk data card defines the stiffness matrix in the material coordinate system for the ply.
   - Ply fiber direction, CORDM field on the PSOLID bulk data card which references a material coordinate system definition.

5. **Assign solid ply properties to the elements** that represent the ply. This process assigns an element its stiffness matrix through the PID field on the CPENTA or CHEXA bulk data cards.

6. **Create boundary conditions** applied to the composite model that simulate the in-situ environments under investigation.

   - Constraints are defined in OptiStruct through the SPC bulk data card.
   - Forces and Moments are defined in OptiStruct through the FORCE and MOMENT bulk data cards respectively.
   - Pressures on solid element faces are defined in OptiStruct through the PLOAD4 bulk data card.
   - Initial and Model temperature distributions are defined in OptiStruct through the TEMP or TEMPD bulk data cards.

7. **Create load steps** for each load case that the composite model is to be analyzed for by combining appropriate boundary conditions that simulate the in-situ environments of the composite structure under investigation. Load steps are defined in OptiStruct through the SUBCASE, ANALYSIS, TITLE, SPC, LOAD, and TEMPERATURE(LOAD) subcase control cards.

8. **Create control cards** to define initial temperatures, output results, output formats, and solver controls.

   - The initial temperature distribution is defined in OptiStruct through the TEMPERATURE(INITIAL) subcase control card.
   - Displacement output is defined in OptiStruct through the DISPLACEMENT i/o options card.
   - Solid ply strain output is defined in OptiStruct through the STRAIN i/o options card. If thermal boundary conditions are applied, then it is important to request mechanical strain output using the MECH option on the STRAIN i/o options card.
   - Solid ply stress output is defined in OptiStruct through the STRESS i/o options card.
   - Output file formats are defined in OptiStruct through the OUPTUT i/o options card. Typically H3D (binary file for post-processing in Altair HyperView) and ASCII (text file) file formats are requested.

9. **Export the solver input file** representing the composite analysis model from the pre-processor (HyperMesh) and solve the composite analysis by submitting the solver input file to the solver executable (OptiStruct).

10. **Post-process the composite analysis results**. The most important results for composite models are the solid ply mechanical strains and stresses in the material coordinate system. It is recommended that users consult the specific solver documentation to determine the exact coordinate system and type of strain (total, mechanical, and/or thermal) that are output from that solver. Most solvers, by default, output solid ply results in the material coordinate system and total strain tensors. Note that the mechanical strain tensors must be used whenever there is a thermal boundary condition applied. If there is no thermal boundary condition applied, then the mechanical strain tensor is equivalent to the total strain tensor and the default output from most solvers can be used directly.

Chapter 10
Exercises

10.1. Create an OptiStruct composite <u>zone-based shell model</u> of the laminated plate given in figure 10.1. The laminate is made from unidirectional plies of thickness t = 0.01" with transversely isotropic material properties as below. Compare the results of this model with those obtained from problem 6.7. Are they the same?

Ply material properties

$E_1 = 22.0e6$ psi
$E_2 = 1.30e6$ psi
$v_{12} = 0.30$
$v_{23} = 0.26$
$G_{12} = 0.75e6$ psi
$G_{23} = 0.516e6$ psi
$\alpha_1 = -0.30e-6$ /°F
$\alpha_2 = 18.0e-6$ /°F

Ply material strengths

$X_T = 170,000$ psi
$X_C = 170,000$ psi
$Y_T = 6,500$ psi
$Y_C = 28,000$ psi
$S = 10,000$ psi

10.2. Create an OptiStruct composite <u>ply-based shell model</u> of the laminated plate as given in figure 10.1. The laminate is made from unidirectional plies of thickness t = 0.01" with transversely isotropic material properties as given in problem 10.1. Compare the results of this model with those obtained from problem 6.7 and 10.1. Are they the same?

10.3. Create an OptiStruct composite ply-by-ply solid model of the laminated plate as given in figure 10.1. The laminate is made from unidirectional plies of thickness t = 0.01" with transversely isotropic material properties as given in problem 10.1. Compare the results of this model with those obtained from problem 6.7, 10.1, and 10.2. Are they the same?

10.4. Create an OptiStruct composite <u>zone-based shell model</u> of the laminated plate shown below. The laminate is made from unidirectional plies of thickness t = 0.01" with transversely isotropic material properties as below. Feel free to mesh the plate with a finer mesh than shown.

Ply material properties

$E_1 = 22.0e6$ psi

$E_2 = 1.30e6$ psi

$v_{12} = 0.30$

$v_{23} = 0.26$

$G_{12} = 0.75e6$ psi

$G_{23} = 0.516e6$ psi

$\alpha_1 = -0.30e{-6}$ /°F

$\alpha_2 = 18.0e{-6}$ /°F

$X_T = 170,000$ psi

$X_C = 170,000$ psi

$Y_T = 6,500$ psi

$Y_C = 28,000$ psi

$S = 10,000$ psi

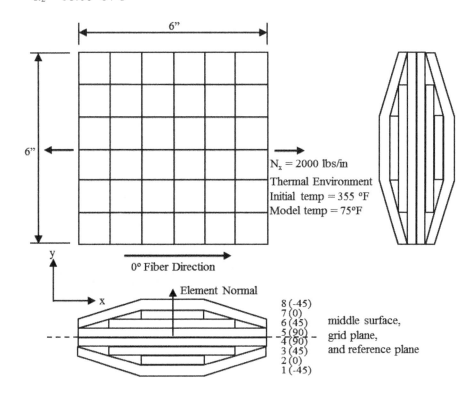

10.5. Create an OptiStruct composite <u>ply-based shell model</u> of the laminated plate given in problem 10.3. The laminate is made from unidirectional plies of thickness (t = 0.01") with transversely isotropic material properties as given in problem 10.3. Feel free to mesh the plate with a finer mesh than shown.

10.6. Modify the OptiStruct composite <u>zone-based shell model of problem 10.4</u> by removing ply 4 and ply 5 from the laminate. How did you have to modify the composite zone-based model to accommodate this design change?

10.7. Modify the OptiStruct composite <u>ply-based shell model of problem 10.5</u> by removing ply 4 and ply 5 from the laminate. How did you have to modify the composite ply-based model to accommodate this design change?

Chapter 11
Composite Design Optimization

The general optimization problem can be mathematically defines as;

Minimize $f(X) = f(X_1, X_2, X_3 ..., X_n)$ (11.1)

subject to;

$g_j(X) \leq 0.0$ j = 1, ...,m   (11.2)

$X_i^L \leq X_i \leq X_i^U$ i = 1, ...,n   (11.3)

where;

f(X) is the objective function to be minimized and $g_j(X)$ are the constraint functions which must be satisfied. Both f(X) and $g_j(X)$ are functions of the design variables X_i, each of which can have lower and upper bound limits. There are m constraints and n design variables for any optimization problem. Design variables are values which can be changed in a model; such as thickness, length, width, and other model parameters. Design variables typically have lower and upper bound limit constraints which must be satisfied for a feasible design. The objective function is a single response of the model that is to be minimized (or maximized). The constraint functions are responses of the model which must be satisfied for a feasible design. The objective function and the constraint functions are both functions of the design variables. Responses are values which are measured from a model; such as volume, mass, compliance (inverse of stiffness), displacement, strain, stress, and other model responses. Exactly one response must be the objective function; all other responses can be constraint functions.

A popular acronym used to help remember optimization setup definitions is DRCO; Design variables, Responses, Constraints, and Objective. Each of which must be specified completely to define an optimization design model given a running analysis model.

D – Design variables are values which can be changed in a model.
R – Responses are values which are measured from a model.
C – Constraints are limits on the responses of the model which must be satisfied for a feasible design.
O – Objective is a single response of the model which is to be minimized (or maximized)

The composite design optimization methodology presented within this book was developed to solve very complex composite design optimization problems given a current design or a "clean sheet". The methodology breaks down the complex design optimization problem, which is not solvable by itself, into several simpler design optimization problems, which are solvable by themselves. The cumulative solution to each of the simpler design optimization problems provides a solution to the complex design optimization problem. This process of breaking down complex problems into several simpler problems is consistent with the engineering method that has been employed successfully for hundreds of years by engineers throughout history. Therefore, the composite design optimization methodology presented herein is a three step design optimization process. Each step provides a solution to the following design optimization questions.

1. **Composite Free-Size Optimization**. This step of the composite design optimization methodology answers the question; "What ply shapes, for each ply layer, would build up the most efficient composite part?"

2. **Composite Size Optimization**. This step of the composite design optimization methodology answers the question; "Exactly many plies of each ply shape are required to satisfy strength and manufacturing engineering requirements?"

3. **Composite Shuffling Optimization**. This step of the composite design optimization methodology answers the question; "What are possible stacking sequences that satisfy final part manufacturing requirements?"

4. **Final Design Analysis**. This step of the composite design optimization methodology verifies that the final design is feasible and meets all constraints.

The remainder of this chapter is devoted to discussing the complete composite design optimization methodology by specifying the general process to develop composite free-size optimization, composite size optimization, and composite shuffling optimization models. The general process for each step of the composite design optimization methodology is described by reference to a composite design optimization problem for an open hole tension specimen as defined in figure 11.1 and 11.2.

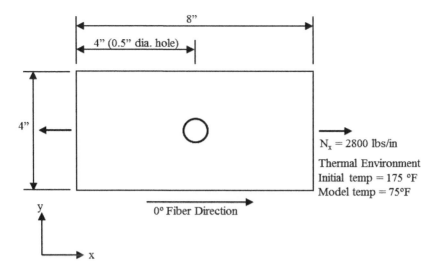

Design Optimization Constraints and Objectives
- Consider 0/90/−45/45 plies
- Ply thickness is 0.01"
- All laminates must be balanced and symmetric
- 0° plies must be between 20% and 65% of the total laminate thickness
- Laminate zones must have a drop off no greater than 1:3
- Fiber strain must not exceed 3000 µε in any ply
- First ply failure must not occur in any ply
- Minimize the mass of the part

Figure 11.1, Open Hole Tension Design Optimization Problem

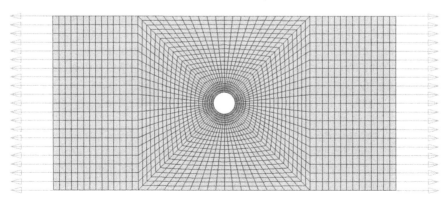

Figure 11.2, Open Hole Tension Model

11.1 Composite Free-Size Optimization

The composite free-size optimization step of the composite design optimization methodology answers the question; "What ply shapes, for each ply layer, would build up the most efficient composite part?" At this stage in the design optimization process the part shape is known. Solid or shell topology could have previously been performed or a part shape could have been given. However, the ply shapes which build up the part thicknesses and define the constant thickness zones (laminate zones) is unknown and needs to be determined through the ply shapes by using composite free-size optimization technology. As can be seen from the initial question, composite free-size optimization is concept design optimization. Therefore, engineering interpretation of the resulting ply shapes is typically required. It is not the objective of a composite free-size optimization to develop a design which meets all the engineering requirements. But instead, it should be used to gain an understanding of what the most efficient structure would desire to be if it could exist with no further constraints, which will be added in subsequent steps.

The general process for defining a composite free-size optimization is described in this section. The required i/o options, subcase control, and bulk data cards necessary to build OptiStruct composite free-size optimization models can be referenced in appendices A, B, C and D respectively. The final OptiStruct composite free-size model can be referenced in appendix F.

1. **Develop a ply-based analysis model** as outlined in section 10.2. The first step in any design optimization problem is to verify a proper running analysis model. Further, ensure that all laminates are defined with SMEAR technology enabled. At this stage the ply fiber directions to consider within the design optimization problem are known, but not how to stack those plies. Furthermore, when a stacking sequence is not specified or known, SMEAR technology should be enabled to make the problem stacking-sequence independent. Laminates are defined in OptiStruct through the STACK bulk data card. SMEAR technology is enabled by setting the LAM field on the STACK card to SMEAR or SYMSMEAR.

2. **Define composite free-size design variables**. The composite free-size design variables are the thickness of every ply for every element. In our example composite design optimization problem, we are considering four plies 0/90/−45/45 with 2,240 elements as defined in the figure 11.2. Therefore, 8,960 thickness design variables will exist in this composite free-size optimization. Unlike topology optimization, in which the optimization algorithm is "pushing" the design variables to either their

lower or upper bound, the free-size optimization algorithm allows the thickness design variables to "freely" be any value between their lower and upper bounds. In this light, the composite free-size optimization algorithm captures the coupling between total element thickness and the relative percentage of each ply's thickness to the total element thickness (i.e. the laminate family). As an example, if the composite free-size optimization algorithm decides it needs to increase an element thickness, it has a choice of ply by which to achieve the increase in element thickness. If it chooses to increase the $0°$ ply thickness, as opposed to the $90°$ ply thickness, the stiffness increase effect due to the increase in element thickness will be significantly amplified in the $0°$ direction by the selection of the $0°$ ply. Therefore, the free-size optimization algorithm is not only optimizing on the total element thicknesses, but also on the relative percentage of each ply's thickness to the total element thickness (i.e. the laminate family). It is important to state again that SMEAR technology should be utilized at the composite free-sizing stage; thus, making the optimization problem stacking-sequence independent. With SMEAR technology, regardless of the stacking sequence or how ply thicknesses grow or shrink (i.e. add or remove plies within that ply layer), the composite free-size results will be the same. Composite free-size design variables are defined in OptiStruct through the DSIZE bulk data card.

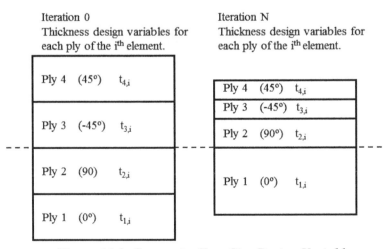

Figure 11.3, Composite Free-Size Design Variables

3. **Define composite manufacturing constraints** for the composite free-size optimization. Composite manufacturing constraints provide a mechanism for the free-size optimization algorithm to produce manufacturable designs and avoid the trivial "all 0 degree" optimized design result. The

composite manufacturing constraints include; total laminate thickness constraints, ply group percentage constraints, ply balancing constraints, ply group constant thickness constraints, and ply group drop off constraints. Each composite manufacturing constraint is defined in figure 11.4 through figure 11.8 below. Composite manufacturing constraints for free-size optimization are defined in OptiStruct through the DSIZE bulk data card, COMP continuation lines.

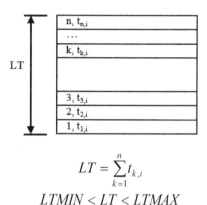

$$LT = \sum_{k=1}^{n} t_{k,i}$$

$$LTMIN < LT < LTMAX$$

Figure 11.4, Total Laminate Thickness Manufacturing Constraint

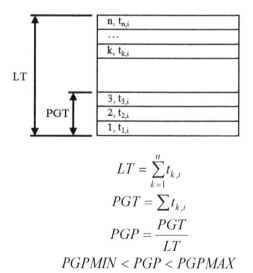

$$LT = \sum_{k=1}^{n} t_{k,i}$$

$$PGT = \sum t_{k,i}$$

$$PGP = \frac{PGT}{LT}$$

$$PGPMIN < PGP < PGPMAX$$

Figure 11.5, Ply Group Percentage Manufacturing Constraint

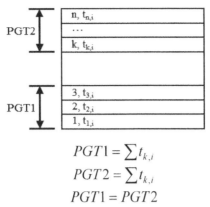

$$PGT1 = \sum t_{k,i}$$
$$PGT2 = \sum t_{k,i}$$
$$PGT1 = PGT2$$

Figure 11.6, Ply Group Balancing Manufacturing Constraint

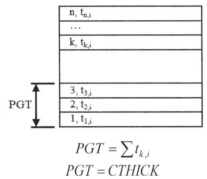

$$PGT = \sum t_{k,i}$$
$$PGT = CTHICK$$

Figure 11.7, Ply Group Constant Thickness Constraint

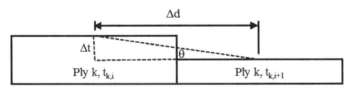

$$PDMAX = \tan(\theta) = \frac{\Delta t}{\Delta d} = \frac{t_{k,i} - t_{k,1+i}}{\Delta d}$$

Figure 11.8, Ply Group Drop Off Constraint

4. **Define responses** for the composite free-size optimization. One of the typical (but not the only) composite free-size optimization problem setups is to minimize the compliance (i.e. maximize the stiffness) while constraining the volume fraction to be less than 30% of the original volume of the structure. Effectively this composite free-size optimization problem setup attempts to "starve" the part of material by specifying that

70% of the initial material volume must be removed while additionally specifying that the remaining 30% of the material should be distributed in such a way as to provide the "stiffest" part possible. This type of composite free-size optimization problem setup tends to place material only where it is absolutely necessary; thus, producing weight efficient concept design results. For this particular composite free-size optimization problem setup, volume fraction and compliance responses are required. Responses are defined in OptiStruct through the DRESP1 bulk data card.

5. **Define constraints** for the composite free-size optimization. As discussed above, the constraint for our particular composite free-size optimization problem definition is that the volume fraction response must be less than 30% of the original volume of the structure. That is to say, 70% of the initial material volume must be removed. Constraints are defined in OptiStruct through the DCONSTR and DCONADD bulk data cards.

6. **Define the objective** for the composite free-size optimization. As discussed above, the objective for our particular composite free-size optimization problem definition is to minimize the compliance. That is to say, maximize the stiffness of the part since compliance is the inverse of stiffness. An objective is defined in OptiStruct through the DESOBJ subcase control card.

7. **Assign constraint sets to each subcase** by defining global or subcase dependent constraint sets. Constraint sets are assigned to subcase definitions in OptiStruct through the DESGLB or DESSUB subcase control cards.

8. **Create control cards** to define additional optimization output results and output formats. Thickness optimization results and automatic free-size to size model output generation should be requested. Thickness optimization results are requested in OptiStruct through the THICKNESS i/o options card. Optimization output formats are defined in OptiStruct through the OUTPUT i/o options card. Use OUTPUT, FSTOSZ to automatically output a size model containing the "sliced" ply results from the free-size model.

9. **Export the solver input file** representing the composite free-size optimization model from the pre-processor (HyperMesh) and solve the composite free-size optimization by submitting the solver input file to the solver (OptiStruct) executable.

10. **Post-process the composite free-size optimization results**. There are two important output files; *_des.h3d and *_sizing.#.fem. The *_des.h3d file contains the composite free-size optimization thickness results which can be contour plotted to facilitate interpretations of the resulting optimized ply shapes. However, the most important output file is the *_sizing.#.fem file. This file contains a "run ready" composite size optimization input file. While this file is "run ready", it is highly suggested to import and modify this model within a pre-processor (HyperMesh) as necessary. The significant advantage to the *_sizing.#.fem file is that optimized ply shapes from the composite free-size optimization are contained in this file; and design variables and design variable property relationships for the thickness of each ply shape are automatically generated. Ply shapes are generated for each ply by "slicing" the composite free-size optimization thicknesses of a single ply for every element as shown in figure 11.8. This process is repeated for every ply and the resulting composite free-size ply shapes are shown in figure 11.9.

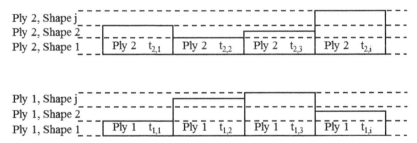

Figure 11.9, Composite Free-Size Ply Shape Generation

The ply numbering convention within OptiStruct is [LPPSNN].

Where;
- L is the laminate number $(1, 2, 3, \dots) \leq 9$
- P is the ply number $(01, 02, 03, \dots) \leq 99$
- S is the ply shape number for the given ply $(1, 2, 3, \dots) \leq 9$
- NN counts the ply iterates for a given ply shape $(01, 02, 03, \dots) \leq 99$

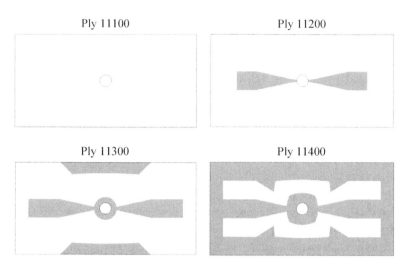

Figure 11.10, Composite Free-Size 0° Ply Shapes
Note: ply shapes are denoted by white boundaries.

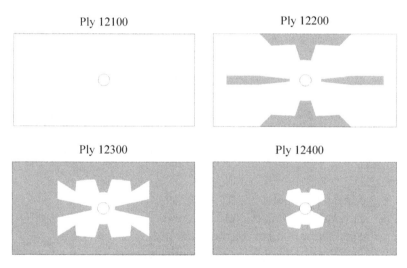

Figure 11.11, Composite Free-Size 90° Ply Shapes
Note: ply shapes are denoted by white boundaries.

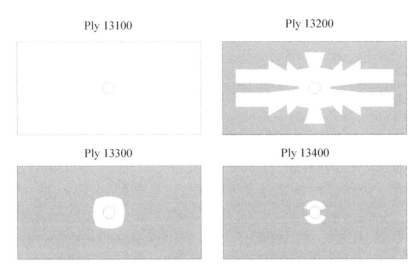

Figure 11.12, Composite Free-Size 45°/-45° Ply Shapes
Note: ply shapes are denoted by white boundaries.

11.2 Composite Size Optimization

This step of the composite design optimization methodology answers the question; "Exactly many plies of each ply shape are required to satisfy strength and manufacturing engineering requirements?" At this stage, both the part shape and the ply shapes define the constant thickness zones are known. However, exactly how many plies of each ply shape that are required to meet strength and manufacturing engineering targets is unknown and needs to be determined through composite size optimization technology. Composite size optimization is detailed design optimization as opposed to composite free-size optimization which is concept design optimization.

The general process for defining a composite size optimization is described in this section. The required i/o options, subcase control, and bulk data cards necessary to build OptiStruct composite size optimization models can be referenced in appendices A, B, C and D respectively. The final OptiStruct composite size model can be referenced in appendix F.

1. **Import the composite free-size *_sizing.#.fem file**. Make sure that all the plies have the same initial thickness, typically 2-4 plies thick. Also make sure that a manufacturing thickness is defined for each ply using the TMANUF field on the PLY bulk data card. This causes discrete ply thicknesses to be selected during the composite size optimization. Finally,

laminates should be defined with symmetric smear technology by setting the LAM field on the STACK bulk data card to SYSSMEAR. Symmetric smear makes the problem stacking-sequence independent and ensures that a symmetric laminate will result by automatically doubling the number of plies.

2. **Define composite size design variables and design variable property relationships.** The *_sizing.#.fem file automatically defines design variables and design variable property relationships for the thickness of each ply shape resulting from the composite free-size optimization. However, it is suggested that the design variable upper and lower bound limits be adjusted appropriately. By default, OUTPUT FSTOSZ produces 4 ply shapes for each ply. Since we are considering 0/90/45/-45 plies in our open hole tension design example, there are a total of 16 design variables. Design variables are defined in OptiStruct through the DESVAR bulk data card. Design variable property relationships are defined in OptiStruct through the DVPREL1 bulk data card.

3. **Define composite manufacturing constraints** for the composite size optimization. The *_sizing.#.fem file automatically defines composite manufacturing constraints as copies of those which were defined in the composite free-size optimization. Composite manufacturing constraints provide a mechanism for the size optimization algorithm to produce manufacturable designs and avoid the trivial "all 0 degree" optimized design result. The composite manufacturing constraints include; total laminate thickness constraints, ply group percentage constraints, ply balancing constraints, ply group constant thickness constraints, and ply group drop off constraints. Each composite manufacturing constraint is defined in figure 11.4 through figure 11.8 above in section 11.1. Composite manufacturing constraints for size optimization are defined in OptiStruct through the DCOMP bulk data card.

4. **Define responses** for the composite size optimization. The responses needed for our particular composite design optimization problem are mass, fiber mechanical strain, and first ply failure using Tsai-Wu first ply failure criterion. Before defining these responses, you should first delete the volume fraction and compliance responses left over from the composite free-size optimization. Responses are defined in OptiStruct through the DRESP1 bulk data card.

5. **Define constraints** for the composite size optimization. The constraints needed for our particular composite design optimization problem are fiber mechanical strain less than 3000 $\mu\varepsilon$ and no first ply failure as defined by the Tsai-Wu first ply failure criterion. Constraints are defined in OptiStruct through the DCONSTR and DCONADD bulk data cards.

6. **Define the objective** for the composite size optimization. The objective for our particular composite design optimization problem is minimize the mass of the structure. An objective is defined in OptiStruct through the DESOBJ case control card.

7. **Assign constraint sets to each subcase** by defining global or subcase dependent constraint sets. Constraint sets are assigned to subcase definitions in OptiStruct through the DESGLB or DESSUB subcase control cards.

8. **Create control cards** to define additional optimization output results and output formats. Thickness optimization results and automatic size to shuffling model output generation should be requested. Thickness optimization results are requested in OptiStruct through the THICKNESS i/o options card. Optimization output formats are defined in OptiStruct through the OUTPUT i/o options card. Use OUTPUT, SZTOSH to automatically output a shuffling model containing the ply thickness results (i.e. # of plies) from the size model.

9. **Export the solver input file** representing the composite size optimization model from the pre-processor and solve the composite size optimization by submitting the solver input file to the solver (OptiStruct) executable.

10. **Post-process the composite size optimization results**. The most important output file is the *_shuffling.#.fem file. This file contains the number of plies of each ply shape which are required to meet the strength and manufacturing engineering targets. Importing the *_shuffling.#.fem into a preprocessor (HyperMesh) shows the final number of plies for each ply shape. In addition the .out file contains the same information for each iteration of the composite size optimization. Reviewing this file after each optimization is a suggested practice. Still, even after a composite size optimization, the final design is not completely defined. The exact stacking sequence for the plies is still unknown and will be determined in the next step, composite shuffling optimization.

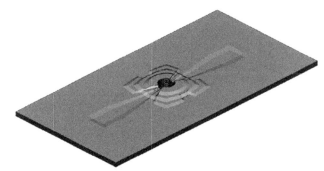

Figure 11.13, Composite Size Optimization Ply Shapes

Table 11.1, Composite Size Optimization Results – Number of Plies

Ply Number	Size Optimization Thickness (in")	Actual Number of Plies (SYM)
11100	0.02	4
11200	0.02	4
11300	0.00	0
11400	0.00	0
12100	0.01	2
12200	0.00	0
12300	0.00	0
12400	0.03	6
13100	0.01	2
13200	0.00	0
13300	0.02	4
13400	0.02	4
14100	0.01	2
14200	0.00	0
14300	0.02	4
14400	0.02	4
Total		36

11.3 Composite Shuffling Optimization

This step of the composite design optimization methodology answers the question; "What are possible stacking sequences that satisfy final part manufacturing requirements?" At this stage many things are known; including the part shape, the ply shapes which define the constant thickness zones, and even the exact number of plies of each ply shape are known. However, exactly how to stack those plies to meet manufacturing engineering requirements is unknown and needs to be determined through composite shuffling optimization.

The general process for defining a composite shuffling optimization is described in this section. The required i/o options, subcase control, and bulk data cards necessary to build OptiStruct composite shuffling optimization models can be referenced in appendices A, B, C and D respectively. The final OptiStruct composite shuffling model can be referenced in appendix F.

1. **Import the composite size *_shuffling.#.fem file**

2. **Add shuffling design variable manufacturing constraints.** Typically the maximum successive number of plies for all ply orientations are limited to four with zero violation. The maximum successive number of plies for a given layer is defined via the MAXSUCC continuation line on the DSHUFFLE bulk data card. A cover stacking sequence is typically defined as [-45/0/45/90] with as many repeats as necessary. The cover stacking sequence defines the plies at the top and bottom surface of the laminate. The cover stacking sequence is defined via the COVER continuation line on the DSHUFFLE bulk data card. Finally a core stacking sequence can be defined with as many repeats as necessary. The core stacking sequence defines the plies at the middle surface of the laminate. The core stacking sequence is defined via the CORE continuation line on the DSHUFFLE bulk data card.

3. **Export the solver input file** representing the composite shuffling optimization model from the pre-processor (HyperMesh) and solve the composite size optimization by submitting the solver input file to the solver (OptiStruct) executable.

4. **Post-process the composite shuffling optimization results**. The most important output file is the *.prop file. This file contains the final stacking sequence which meets the strength and manufacturing engineering targets. Importing the *.prop with FE-overwrite into a preprocessor (HyperMesh) will produce the final verification model. In addition the *.shuffle.html file contains information on the stacking sequence as the shuffling optimization iterations progressed to the final stacking sequence. This file can provide useful information and it is suggested practice to review the file contents. However, the final design is not complete. The final design must be verified to meet the desired strength and manufacturing engineering targets. This last step will be discussed in the next section.

Table 11.2, Composite Shuffle Optimization Results – Final Stacking Sequence

Ply	Thickness	Fiber Orientation	Shape
14101	0.01	−45	Full
11101	0.01	0	Full
13101	0.01	45	Full
12101	0.01	90	Full
14301	0.01	-45	Partial
11201	0.01	0	Partial
13301	0.01	45	Partial
12401	0.01	90	Partial
14302	0.01	-45	Partial
11202	0.01	0	Partial
13302	0.01	45	Partial
12402	0.01	90	Partial
14401	0.01	-45	Partial
11102	0.01	0	Full
13401	0.01	45	Partial
12403	0.01	90	Partial
14402	0.01	-45	Partial
13402	0.01	45	Partial
13403	0.01	45	Partial
14403	0.01	-45	Partial
12404	0.01	90	Partial
13404	0.01	45	Partial
11103	0.01	0	Full
14404	0.01	-45	Partial
12405	0.01	90	Partial
13303	0.01	45	Partial
11203	0.01	0	Partial
14303	0.01	-45	Partial
12406	0.01	90	Partial
13304	0.01	45	Partial
11204	0.01	0	Partial
14304	0.01	-45	Partial
12102	0.01	90	Full
13102	0.01	45	Full
11104	0.01	0	Full
14102	0.01	-45	Full

11.4 Final Design Verification

This step of the composite design optimization methodology verifies the final design meets the strength and manufacturing engineering requirements by performing an analysis on the final design as given from the results of the shuffling optimization. In this design example we had a fiber strain limit of 3000 με. Figure 11.14 show a contour plot of maximum mechanical strain in the normal x-axis direction (i.e. fiber direction) for all plies at the design load. From this contour the maximum fiber strain is 2400 με which is less than our 3000 με limit. Also in this design example we had a requirement that first ply failure not exist in any ply. Figure 11.15 shows a contour plot of the maximum Tsai-Wu failure index for all plies at the design load. From this contour the maximum failure index is 0.77 which is less than our 1.0 limit. Therefore all strength requirements are meet in the final design. The final design mass is 0.258 lbs. The final design OptiStruct model can be referenced in appendix F.

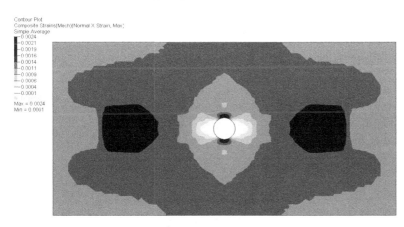

Figure 11.14, Fiber Direction Mechanical Strain at Design Load

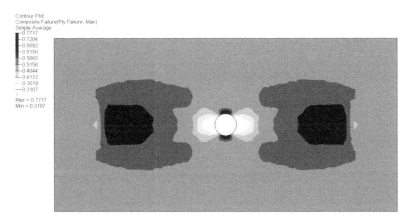

Figure 11.15, First Ply Failure Index at Design Load

Chapter 11
Exercises

11.1. Create a minimum mass composite design for the open-hole tension model as given in figure 11.1. The optimized laminate can be made from 0/90/−45/45 unidirectional plies of thickness (t = 0.01") with transversely isotropic material properties as given below. The optimized laminated must be balanced and symmetric; and the 0/90 plies must be between 20% and 65% of the total laminate thickness. Also, all laminate zones must have a drop off no greater than 1:3. Finally, the fiber strain must not exceed 3000 $\mu\varepsilon$ and no first ply failure can occur in any ply. (Hint: a more refined mesh may be necessary to obtain a converged optimal design than what is presented in this chapter.)

Ply material properties
$E_1 = 22.0e6$ psi
$E_2 = 1.30e6$ psi
$\nu_{12} = 0.30$
$\nu_{23} = 0.26$
$G_{12} = 0.75e6$ psi
$G_{23} = 0.516e6$ psi
$\alpha_1 = -0.30e{-}6$ /°F
$\alpha_2 = 18.0e{-}6$ /°F

Ply material strengths
$X_T = 170,000$ psi
$X_C = 170,000$ psi
$Y_T = 6,500$ psi
$Y_C = 28,000$ psi
$S = 10,000$ psi

Appendix A
OptiStruct I/O Options Reference

CSTRAIN

Defines composite ply strain output. Composite ply strains are output at the middle of each ply.

CSTRAIN (FORMAT, TYPE, EXTRA) = OPTION

Argument	Description
FORMAT	Defines the output format. (Default = blank)

        ASCII - Results are output to the ASCII file formats.
        H3D - Results are output to the *.h3d file format.
        OP2 - Results are output to the *.op2 file format.
        Blank - Results are output in all active formats defined on the OUTPUT card.

TYPE   Defines the strain components to output. (Default = ALL)

        ALL – All strain components and principals are output.
        PRINC – Only principal strains are output.

EXTRA   Defines extra parameters for composite strain output.

        MECH - Total and mechanical strains are output.
        THERM - Total and thermal strains are output.

OPTION   Defines the output options. (Default = ALL)

        ALL – Results are output for all elements.
        NONE – Results are not output.
        SID – Results are output for all elements defined by the element set identification number.
        PID – Results are output for all elements that reference the properties defined by the property set identification number.

CSTRESS

Defines composite ply stress output. Composite ply stresses are output at the middle of each ply.

CSTRESS (FORMAT, TYPE) = OPTION

Argument	Description
FORMAT	Defines the output format. (Default = blank)

ASCII - Results are output to the ASCII file formats.
H3D - Results are output to the *.h3d file format.
OP2 - Results are output to the *.op2 file format.
Blank - Output in all formats defined on the OUTPUT card.

TYPE   Defines the stress components to output. (Default = ALL)

ALL – All stress components, principals, and failure indices are output.
PRINC – Only principal stresses are output.
FI – Only failure indices are output.

OPTION   Defines the output options. (Default = ALL)

ALL – Results are output for all elements.
NONE – Results are not output.
SID – Results are output for all elements defined by the element set identification number.
PID – Results are output for all elements that reference the properties defined by the property set identification number.

DISPLACEMENT

Defines grid point displacement output.

DISPLACEMENT (FORMAT) = OPTION

Argument	Description
FORMAT	Defines the output format. (Default = blank)

ASCII - Results are output to the ASCII file formats.
H3D - Results are output to the *.h3d file format.
OP2 - Results are output to the *.op2 file format.
Blank - Results are output in all active formats defined on the OUTPUT card.

OPTION   Defines the output options. (Default = ALL)

ALL – Results are output for all grids.
NONE – Results are not output.
SID – Results are output for all grids defined by the node set identification number.

OUTPUT

Defines active output formats for an analysis or optimization run.

OUTPUT, FORMAT, FREQUENCY

Argument	Description
FORMAT	Defines the output format.

Analysis Results Output Formats
H3D - Results are output to the binary *.h3d file format.
OP2 - Results are output to the binary *.op2 file format.
ASCII - Results are output to the OptiStruct ASCII file formats (*.cstr composite stress/strain, *.disp displacement, *.force, *.gpf grid point forces, *.load applied load, *.mpcf multi-point constraint forces, *.spcf single point constraint forces, *.strs stress/strain)
PUNCH - Results are output to the Nastran *.pch file format.

Optimization Results Output Formats
FSTOSZ - Automatic generation of a size optimization model at the last iteration of the free-size optimization. (*_sizing.#.fem)
SZTOSH - Automatic generation of a shuffling optimization model at the last iteration of the size optimization model. (*_shuffling.#.fem)
DESVAR - Outputs the updated design variables for the given iteration to the *.desvar and/or *.out files.
PROPERTY - Outputs the updated property cards for the given iteration to the *.prop and/or *.out files.

FREQUENCY Defines the frequency at which results are output. (Default = FL)

FIRST - Results are output for the first iteration only.
LAST - Results are output for the last iteration only.
FL - Results are output for the first and last iterations.
ALL - Results are output for all iterations.
N= - Results are output for the Nth iteration.
NONE - Results are not output.

STRAIN

Defines homogeneous strain output. For shell elements, homogeneous strains are output at Z1 and Z2 locations as defined on the PSHELL card. (Default $Z1 = -T/2$; Default $Z2 = T/2$)

STRAIN (FORMAT, TYPE, LOCATION, EXTRA) = OPTION

Argument	Description
FORMAT	Defines the output format. (Default = blank)

ASCII - Results are output to the ASCII file formats.
H3D - Results are output to the *.h3d file format.
OP2 - Results are output to the *.op2 file format.
Blank - Output in all formats defined on the OUTPUT card.

TYPE   Defines the strain components to output. (Default = ALL)

VON - Only von Mises strain is output.
PRINC - Only Principal and von Mises strain are output.
ALL - All strain components are output.

LOCATION   Defines the locations within shell and solid elements where the strain components are output. (Default = CENTER)

CENTER - Output at the element center only.
CUBIC - Output at the element center and grid points using the strain gage approach with cubic bending correction.
CORNER and **BILIN** - Output at the element center and the grid points with bilinear extrapolation.

EXTRA   Defines extra parameters for homogeneous strain output.

MECH - Total and mechanical strains are output.
THERM - Total and thermal strains are output.

OPTION   Defines the output options. (Default = ALL)

ALL – Results are output for all elements.
NONE – Results are not output.
SID – Results are output for all elements defined by the element set identification number.
PID – Results are output for all elements that reference the properties defined by the property set identification number.

STRESS

Defines homogeneous stress output. For shell elements, homogeneous stresses are output at Z1 and Z2 locations as defined on the PSHELL card. (Default $Z1 = -T/2$; Default $Z2 = T/2$)

STRESS (FORMAT, TYPE, LOCATION) = OPTION

Argument	Description
FORMAT	Defines the output format. (Default = blank)

ASCII - Results are output to the ASCII file formats.
H3D - Results are output to the *.h3d file format.
OP2 - Results are output to the *.op2 file format.
Blank - Output to all formats defined on the OUTPUT card.

TYPE   Defines the stress components to output. (Default = ALL)

VON - Only the von Mises stress is output.
PRINC - Only principal and invariant stresses are output.
ALL - All stress components, principals, and invariants are output.

LOCATION   Defines the locations within shell and solid elements where the stress components are output. (Default = CENTER)

CENTER - Output at the element center only.
CUBIC - Output at the element center and grid points using the strain gage approach with cubic bending correction.
CORNER and **BILIN** - Output at the element center and the grid points with bilinear extrapolation.

OPTION   Defines the output options. (Default = ALL)

ALL – Results are output for all elements.
NONE – Results are not output.
SID – Results are output for all elements defined by the element set identification number.
PID – Results are output for all elements that reference the properties defined by the property set identification number.

THICKNESS

Defines element and ply thickness output for topology, free-size, and size design optimizations.

THICKNESS (FORMAT, COMP) = OPTION

Argument	Description
FORMAT	Defines the output format. (Default = blank)

ASCII - Results are output to the ASCII file formats.
H3D - Results are output to the *.h3d file format.
OP2 - Results are output to the *.op2 file format.
Blank - Output in all formats defined on the OUTPUT card.

COMP   Defines composite ply thickness output options. (Default = DESIGN)

ALL - Outputs the thickness of all plies
DESIGN - Only outputs the thickness of designable plies.
NOPLY - No ply thickness is output.

OPTION   Defines the output options. (Default = ALL)

ALL – Results are output for all elements.
NONE – Results are not output.

TITLE

Defines a title for the model.

TITLE = NAME

Argument	Description
NAME	Title for the model which is printed in the output files. (String less than 80 characters)

Appendix B
OptiStruct Subcase Control Reference

ANALYSIS

Defines the analysis solution sequence for a subcase. Must be defined within a subcase definition.

ANALYSIS = OPTION

Argument | Description
OPTION | Defines the analysis solution sequence for a subcase.

STATICS - Linear static analysis
NLSTAT - Nonlinear quasi-static analysis
MODES - Normal modes analysis
BUCK - Linear buckling modes analysis

DESGLB

Defines a global design constraint set to apply to all subcases. Must be defined before the first subcase definition and must be subcase independent.

DESGLB = ID

Argument	Description
ID	Identification number of a DCONSTR or DCONADD defining the global design constraint set to apply to all subcases.

DESOBJ

Defines a design objective by selecting a single design response as the design objective for an optimization. If the selected design response is subcase independent, then DESOBJ must be defined before the first subcase definition. If the selected design response is subcase dependent, then DESOBJ must be defined within a subcase definition of the appropriate analysis solution sequence.

DESOBJ (TYPE) = ID

Argument	Description
TYPE	The type of design objective. (Default = MIN)
	MIN - The design objective is to minimize the response.
	MAX - The design objective is to maximize the response.
ID	Identification number of a DRESP1 defining the design response as the design objective for an optimization.

DESSUB

Defines a design constraint set that is subcase dependent. Must be defined within a subcase definition of the appropriate analysis solution sequence.

DESSUB = ID

Argument	Description
ID	Identification number of a DCONSTR or DCONADD defining the design constraint set to apply to the subcase.

LABEL

Defines a label for a subcase. Must be defined within a subcase definition of the appropriate analysis solution sequence.

LABEL = NAME

Argument	Description
NAME	String defining a label for the subcase.

LOAD

Defines a static load set for a subcase. Must be defined within a subcase definition of the appropriate analysis solution sequence.

LOAD = ID

Argument   Description
ID   Identification number of a LOAD, FORCE, MOMENT, PLOAD, GRAV, RFORCE, or SPCD defining the load set to apply to the subcase.

SPC

Defines a single point constraint set for a subcase. Can be defined within a subcase definition of the appropriate analysis solution sequence. If defined before the first subcase definition, then the single point constraint set applies to all subcases definitions.

SPC = ID

Argument	Description
ID	Identification number of a SPCADD or SPC defining the single point constraint set to apply to the subcase.

SUBCASE

Keyword used to indicate the start of a subcase definition. Each subcase definition starts with a new subcase keyword.

SUBCASE = ID

Argument	Description
ID	Subcase identification number.

TEMPERATURE

Defines an initial temperature set, a load temperature set, and/or a material temperature set for a subcase.

TEMPERATURE (TYPE) = ID

Argument	Description
TYPE	The type of temperature set. (Default = BOTH)

INITIAL - The referenced temperature set is used to define the initial temperature distribution for the equivalent static load calculation.

MATERIAL - The referenced temperature set is used to define the material temperature distribution used to determine temperature-dependent material properties on MATTi entries.

LOAD - The referenced temperature set is used to define the load temperature distribution for the equivalent static load calculation.

BOTH - The referenced temperature set is used to define both the load and material temperature distributions.

ID   Identification number of a TEMP or TEMPD defining the temperature set to apply to the subcase.

Appendix C
OptiStruct Analysis Bulk Data Reference

CTRIA3

Defines a triangular shell element with 3 grid points.

(1)	(2)	(3)	(4)	(5)	(6)	(7)	(8)	(9)	(10)
CTRIA3	EID	PID	G1	G2	G3	θ	ZOFFS		
			T1	T2	T3				

CQUAD4

Defines a quadrilateral shell element with 4 grid points.

(1)	(2)	(3)	(4)	(5)	(6)	(7)	(8)	(9)	(10)
CQUAD4	EID	PID	G1	G2	G3	G4	θ	ZOFFS	
			T1	T2	T3	T4			

Field	Comments
EID	Element identification number.
PID	Property identification number of the element. Must reference a PSHELL, PCOMP/G, or PCOMPP card.
Gi	GRID point identification numbers of the element. The order of the grid points defines the element normal direction. See figures below for element normal conventions.
θ	Element material coordinate system orientation angle, in degrees, measured from the vector G1-G2 about the element normal. See figures below for θ conventions.
ZOFFS	Offset from the element grid point plane to the reference plane of the plate element. Overrides the ZOFFS field on the PSHELL card. See figures below for ZOFFS conventions.
Ti	Total thickness of the element at each grid point. Overrides the T field on the PSHELL card. Not used for composite zone based modeling methods using PCOMP/G or composite ply based modeling methods using PCOMPP.

Element material coordinate system, TRIA3

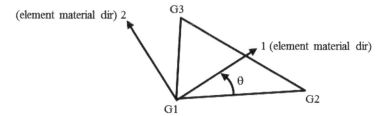

Element coordinate system, TRIA3

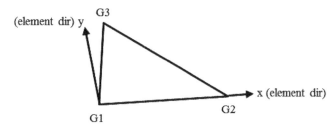

Element material coordinate system, QUAD4

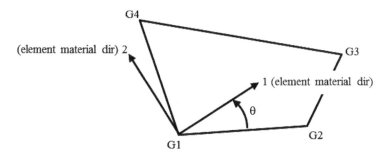

Element coordinate system, QUAD4

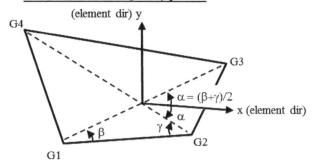

Element normal

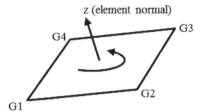

z (element normal)

G4   G3

G2

G1

Element offset

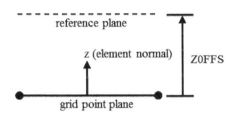

reference plane

z (element normal)   Z0FFS

grid point plane

CPENTA

Defines a pentahedral solid element with 6 grid points.

(1)	(2)	(3)	(4)	(5)	(6)	(7)	(8)	(9)	(10)
CPENTA	EID	PID	G1	G2	G3	G4	G5	G6	

CHEXA

Defines a hexahedral solid element with 8 grid points.

(1)	(2)	(3)	(4)	(5)	(6)	(7)	(8)	(9)	(10)
CHEXA	EID	PID	G1	G2	G3	G4	G5	G6	
	G7	G8							

Field	Comments
EID	Element identification number.
PID	Property identification number of the element. Must reference a PSOLID card.
Gi	GRID point identification numbers of the element. The order of the grid points defines the element topology. See figures below for element topology conventions.

NOTE: Stress and Strain are always output in the material coordinate system as defined on the CORDM field of the PSOLID card assigned to a solid element via the PID field.

CPENTA element topology

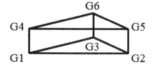

CHEXA element topology

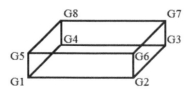

CORD1C / CORD2C

Defines a cylindrical coordinate system with three points.

(1)	(2)	(3)	(4)	(5)	(6)	(7)	(8)	(9)	(10)
CORD1C	CID	G_A	G_B	G_C					
CORD2C	CID	RID	A_x	A_y	A_z	B_x	B_y	B_z	
	C_x	C_y	C_z						

CORD1R / CORD2R

Defines a rectangular coordinate system with three points.

(1)	(2)	(3)	(4)	(5)	(6)	(7)	(8)	(9)	(10)
CORD1R	CID	G_A	G_B	G_C					
CORD2R	CID	RID	A_x	A_y	A_z	B_x	B_y	B_z	
	C_x	C_y	C_z						

Field	Comments
CID	Coordinate system identification number.

RID — Reference coordinate system identification number (default = 0, the basic coordinate system)

G_A, G_B, G_C — Grid point identification numbers defining points A, B, and C.

A_x, A_y, A_z — Coordinates of point A in the reference coordinate system which defines the origin of the coordinate system.

B_x, B_y, B_z — Coordinates of point B in the reference coordinate system which defines the z-axis of the coordinate system as shown in the image below.

C_x, C_y, C_z — Coordinates of point C in the reference coordinate system which defines the rz-plane (CORD2C) or xz-plane (CORD2R) of the coordinate system as shown in the image below.

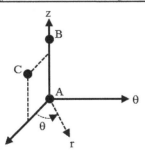

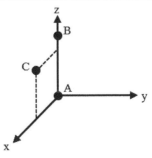

DRAPE

Defines the draping data of a ply in composite ply based modeling.

(1)	(2)	(3)	(4)	(5)	(6)	(7)	(8)	(9)	(10)
DRAPE	ID								
	$DTYPE_1$	DID_1	TF_1	θ_1					
	$DTYPE_2$	DID_2	TF_2	θ_2					
	...								

Field   Comments

ID   Drape data table identification number.

$DTYPE_i$ Row entity type.

             ELEM – Specifics that the row data to follow applies to a single element defined an element ID.
             SET – Specifics that the row data to follow applies to all elements defined an element set ID.

DID_i Row entity type identification number. If $DTYPE_i$ = ELEM, DID_i must refer to an element ID. If $DTYPE_i$ = SET, DID_i must refer to an element set ID.

TF_i Nominal ply thickness factor to achieve the actual ply thickness at the centroid of the element specified. (Default = 1.0)

θ_i Ply nominal fiber direction variation, in degrees, to achieve the ply actual fiber direction at the centroid of the element specified. See figure below for θ_i conventions. (Default = 0.0)

Ply Fiber Orientation Angle

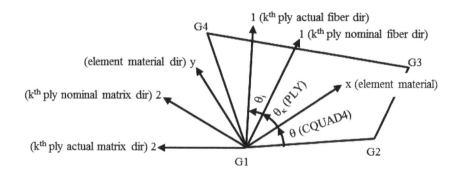

FORCE

Defines a static force at a grid point by specifying a force vector.

(1)	(2)	(3)	(4)	(5)	(6)	(7)	(8)	(9)	(10)
FORCE	SID	GID	CID	F	V_x	V_y	V_z		

Field	Comments
SID	Load set identification number.

GID   Grid point identification number.

CID   Coordinate system identification number (Default = 0)

F   Force vector scale factor.

V_x, V_y, V_z Component of the force vector in the CID coordinate system.

GRID

Defines a grid point.

(1)	(2)	(3)	(4)	(5)	(6)	(7)	(8)	(9)	(10)
GRID	ID	CP	X1	X2	X3	CD	PS		

Field	Comments

ID — Grid point identification number.

CP — Grid point position coordinate system identification number. Defines the coordinate system the grid point coordinate values are entered. (Default = 0, the basic coordinate system)

X1 — Grid point coordinate value in the CP system.

X if CP is a rectangular system
R if CP is a cylindrical or spherical system

X2 — Grid point y-coordinate value in the CP system.

Y if CP is a rectangular system
θ if CP is a cylindrical or spherical system

X3 — Grid point z-coordinate value in the CP system.

Z if CP is a rectangular or cylindrical system
ϕ if CP is a spherical system

CD — Grid point displacement coordinate system identification number. Defines the coordinate system the grid point displacements are output and the single point constraints are entered. (Default = 0, the basic coordinate system)

PS — Grid point permanent single point constraint. Up to six unique digits; 1 through 6.

1 – Translation in the X1 direction
2 – Translation in the X2 direction
3 – Translation in the X3 direction
4 – Rotation about the X1 axis
5 – Rotation about the X2 axis
6 – Rotation about the X3 axis

MAT1

Defines a linear elastic, temperature independent, isotropic material definition for rod, beam, shell, and solid elements.

(1)	(2)	(3)	(4)	(5)	(6)	(7)	(8)	(9)	(10)
MAT1	MID	E	G	ν	ρ	α	TREF	GE	
	ST	SC	SS						

Field	Comments
MID	Material identification number.

E   Young's modulus of the material.

G   Shear modulus of the material.

ν Poisson's ratio of the material.

ρ Mass density of the material.

α Coefficient of thermal expansion of the material.

TREF   Reference stress free temperature of the material. Can be overridden by the TREF field on the PCOMP/G or PCOMPP cards. Typically not used.

GE   Damping coefficient of the material.

ST   Tension stress allowable of the material.

SC   Compression stress allowable of the material.

SS   Shear stress allowable of the material.

Note: Stress allowables are used in laminated plate failure theory calculations if the FT field is specified on the PCOMP/G or PCOMPP card.

MAT2

Defines a linear elastic, temperature independent, anisotropic material definition for shell elements.

(1)	(2)	(3)	(4)	(5)	(6)	(7)	(8)	(9)	(10)
MAT2	MID	Q_{11}	Q_{12}	Q_{13}	Q_{22}	Q_{23}	Q_{33}	ρ	
	α_1	α_2	α_3	TREF	GE	ST	SC	SS	

Field	Comments
MID	Material identification number.

Q_{ij} Components of the plane stress stiffness matrix in the material coordinate system. See Chapter 3 for details on calculation of the plane stress stiffness matrix for a given linear elastic material law.

ρ Mass density of the material.

α_i Coefficient of thermal expansion of the material in the 1-, 2-, and 3-directions.

TREF   Reference stress free temperature of the material. Can be overridden by the TREF field on the PCOMP/G or PCOMPP cards. Typically not used.

GE   Damping coefficient of the material.

ST   Tension stress allowable of the material.

SC   Compression stress allowable of the material.

SS   Shear stress allowable of the material.

Note: Stress allowables are used in laminated plate failure theory calculations if the FT field is specified on the PCOMP/G or PCOMPP card.

MAT8

Defines a linear elastic, temperature independent, orthotropic material definition for shell elements.

(1)	(2)	(3)	(4)	(5)	(6)	(7)	(8)	(9)	(10)
MAT8	MID	E_1	E_2	v_{12}	G_{12}	G_{13}	G_{23}	ρ	
	α_1	α_2	TREF	Xt	Xc	Yt	Yc	S	
	GE	F_{12}	STRN		...				

Field	Comments
MID	Material identification number.

E_1 — Modulus of elasticity of the material in the 1-direction (fiber).

E_2 — Modulus of elasticity of the material in the 2-direction (matrix).

v_{12} — Poisson's ratio of the material on the 1-plane in the 2-direction.

G_{12} — Shear modulus of the material on the 1-plane in the 2-direction.

G_{13} — Shear modulus of the material on the 1-plane in the 3-direction.

G_{23} — Shear modulus of the material on the 2-plane in the 3-direction.

ρ — Mass density of the material.

α_i — Coefficient of thermal expansion of the material in the 1-direction (fiber) and 2-direction (matrix).

TREF — Reference stress free temperature of the material. Can be overridden by the TREF field on the PCOMP/G or PCOMPP cards. Typically not used.

Xt — Tension stress or strain allowable of the material in the 1-direction (fiber).

Xc — Compression stress or strain allowable of the material in the 1-direction (fiber).

Yt — Tension stress or strain allowable of the material in the 2-direction (matrix).

Yc — Compression stress or strain allowable of the material in the 2-direction (matrix).

S   In-Plane shear stress or shear strain allowable of the material.

GE   Damping coefficient of the material.

F_{12} Tsai-Wu interaction term. Used only if FT field is set to TSAI.

STRN   Indicates if Xt, Xc, Yt, Yc, and S fields are entered as stress allowables (0 or blank) or strain allowables (1). (Default = blank).

MAT9

Defines a linear elastic, temperature independent, anisotropic material definition for solid elements.

(1)	(2)	(3)	(4)	(5)	(6)	(7)	(8)	(9)	(10)
MAT9	MID	C_{11}	C_{12}	C_{13}	C_{14}	C_{15}	C_{16}	C_{22}	
	C_{23}	C_{24}	C_{25}	C_{26}	C_{33}	C_{31}	C_{35}	C_{36}	
	C_{44}	C_{45}	C_{46}	C_{55}	C_{56}	C_{66}	ρ	α_1	
	α_2	α_3	α_4	α_5	α_6	TREF	GE		

Field	Comments
MID	Material identification number.

C_{ij} Stiffness matrix coefficients of the material as defined in chapter 3.

ρ Mass density of the material.

α_i Coefficient of thermal expansion vector of the material as defined in chapter 3.

TREF   Reference stress free initial temperature of the material. Overridden by the TEMPERATURE(INITIAL) card in the subcase control section. Typically not used.

GE   Damping coefficient of the material.

MAT9ORT

Defines a linear elastic, temperature independent, orthotropic material definition for solid elements.

(1)	(2)	(3)	(4)	(5)	(6)	(7)	(8)	(9)	(10)
MAT9ORT	MID	E_1	E_2	E_3	v_{12}	v_{23}	v_{31}	ρ	
	G_{12}	G_{23}	G_{31}	α_1	α_2	α_1	TREF	GE	

Field   Comments

MID   Material identification number.

E_1 Modulus of elasticity of the material in the 1-direction (fiber).

E_2 Modulus of elasticity of the material in the 2-direction (matrix).

E_3 Modulus of elasticity of the material in the 3-direction.

v_{12} Poisson's ratio of the material on the 1-plane in the 2-direction.

v_{23} Poisson's ratio of the material on the 2-plane in the 3-direction.

v_{31} Poisson's ratio of the material on the 3-plane in the 1-direction.

G_{12} Shear modulus of the material on the 1-plane in the 2-direction.

G_{23} Shear modulus of the material on the 2-plane in the 3-direction.

G_{31} Shear modulus of the material on the 1-plane in the 3-direction. Equivalent to G_{13}.

ρ Mass density of the material.

α_1 Coefficient of thermal expansion of the material in the 1-direction.

α_2 Coefficient of thermal expansion of the material in the 2-direction.

α_3 Coefficient of thermal expansion of the material in the 3-direction.

TREF   Reference stress free initial temperature of the material. Overridden by the TEMPERATURE(INITIAL) card in the subcase control section. Typically not used.

GE   Damping coefficient of the material.

MOMENT

Defines a static moment at a grid point by specifying a moment vector.

(1)	(2)	(3)	(4)	(5)	(6)	(7)	(8)	(9)	(10)
MOMENT	SID	GID	CID	M	V_x	V_y	V_z		

Field	Comments
SID	Load set identification number.
GID	Grid point identification number.
CID	Coordinate system identification number (Default = 0)
M	Moment vector scale factor.
V_x, V_y, V_z	Component of the moment vector in the CID coordinate system.

PCOMP

Defines the property definition of a laminated plate for composite zone-based shell modeling.

(1)	(2)	(3)	(4)	(5)	(6)	(7)	(8)	(9)	(10)
PCOMP	PID	Z0	NSM	SB	FT	TREF	GE	LAM	
	MID_1	t_1	θ_1	$SOUT_1$	MID_2	t_2	θ_2	$SOUT_2$	
	MID_3	t_3	θ_3	$SOUT_3$	MID_4	t_4	θ_4	$SOUT_4$	
	...								

PCOMPG

Defines the property definition of a laminated plate with global ply identification for composite zone-based shell modeling.

(1)	(2)	(3)	(4)	(5)	(6)	(7)	(8)	(9)	(10)
PCOMPG	PID	Z0	NSM	SB	FT	TREF	GE	LAM	
	$GPLYID_1$	MID_1	t_1	θ_1	$SOUT_1$				
	$GPLYID_2$	MID_2	t_2	θ_2	$SOUT_2$				
	$GPLYID_3$	MID_3	t_3	θ_3	$SOUT_3$				
	...								

PCOMPP

Defines the property definition of a laminated plate for composite ply based modeling.

(1)	(2)	(3)	(4)	(5)	(6)	(7)	(8)	(9)	(10)
PCOMPP	PID	Z0	NSM	SB	FT	TREF	GE		

Field	Comments
PID	Property identification number.

ZO   Distance from the element reference plane to the bottom ply of the laminated plate. See figures below for Z0 conventions. (Default: $-T/2$)

NSM   Nonstructural mass per unit area applied to the laminated plate.

SB   Interlaminate shear stress allowable of the laminated plate.

FT   Laminate failure output option. If blank, no failure output calculations are performed on the laminate. In addition, material allowable fields on the ply MAT cards and the SB field must be entered to perform failure output calculations. (Default: blank)

HILL - Tsai-Hill failure theory.
HOFF - Hoffman failure theory.
TSAI - Tsai-Wu failure theory.
STRN - Max Strain failure theory.

TREF   Reference stress free temperature of the laminated plate. Overrides the TREF field on the MAT card referenced by each ply. If TREF is not specified, then each TREF field on the MAT card referenced by each ply must have the same TREF value.

GE   Damping coefficient of the laminated plate.

LAM   Laminate stacking sequence option. If blank all plies must be specified. (Default = blank)

SYM - Only plies on the bottom half of the laminate need to be specified. This option is not valid for PCOMPG card.
MEM - All plies must be specified, however only [A] matrix terms are calculated. Therefore, the laminated plate exhibits extension behavior only. Any Z0 entry is ignored and set to the default value $(-T/2)$.
BEND - All plies must be specified, however only [D] matrix terms are calculated. Therefore, the laminated plate exhibits bending behavior only. Any Z0 entry is ignored and set to the default value $(-T/2)$.
SMEAR - All plies must be specified and SMEAR technology is utilized to calculate the ABD matrix of the laminate. Any Z0 entry is ignored and set to the default value $(-T/2)$. See chapter 8 for details on SMEAR technology.
SMEARZ0 - All plies must be specified and SMEAR technology is utilized to calculate the ABD matrix of the laminate. The Z0 entry is considered in the calculation of the ABD matrix. Unlike SMEAR technology, SMEARZ0 will develop a B matrix due to the Z0 term. If Z0 is set to the default value $(-T/2)$, then SMEAR and SMEARZ0 will produce the same ABD matrix. See chapter 8 for details on SMEAR technology.
SMCORE - All plies must be specified. The last ply specified must be the core layer. All other plies define the "top" and "bottom" face sheet laminates. Half of the total thickness of the laminate is placed on the "top" of the core. The other half of the laminate thickness is placed on the "bottom" of the core. SMEAR Core technology is utilized to calculate the ABD matrix of the

laminate. Any Z0 entry is ignored and set to the default value (−T/2). See chapter 8 for details on SMEAR Core technology.

SYMEM - Only plies on the bottom half of laminate need to be specified, however only [A] matrix terms are calculated. Therefore, the laminated plate exhibits extension behavior only. Any Z0 entry is ignored and set to the default value (−T/2). This option is not valid for PCOMPG card.

SYBEND - Only plies on the bottom half of the laminate need to be specified, however only [D] matrix terms are calculated. Therefore, the laminated plate exhibits bending behavior only. Any Z0 entry is ignored and set to the default value (−T/2).This option is not valid for PCOMPG card.

SYSMEAR - Only plies on the bottom half of the laminate need to be specified and SMEAR technology is utilized to calculate the ABD matrix of the laminate. See chapter 8 for details on SMEAR technology.

$GPLYID_k$ Global ply identification number of the kth ply. Must be unique with respect to all other plies defined on the current PCOMP/G card.

MID_k Material identification number of the kth ply. Must refer to a MAT1, MAT2, or MAT8 card. If MID_k is not specified for a ply, then the default is the last defined ply's MID_k.

t_k Nominal thickness of the kth ply. If t_k is not specified for a ply, then the default is the last defined ply's t_k.

θ_k Nominal fiber orientation angle, in degrees, of the kth ply relative to the x-axis of the element material coordinate system. See figures below for θ_k conventions.

$SOUT_k$ Stress, strain, and failure output option of the kth ply. Ply stress, strain, and failure output is given at the middle of each ply. In addition, OUTPUT CSTRESS and/or OUTPUT CSTRAIN cards must be defined in the I/O section to get output for the ply. (Default = NO)

NO – do not output stress, strain, or failure for the kth ply.
YES – output stress, strain, and failure for the kth ply.

Laminate Stacking Sequence and Offset Definitions

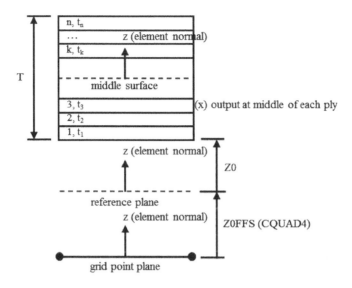

Ply Fiber Orientation Angle

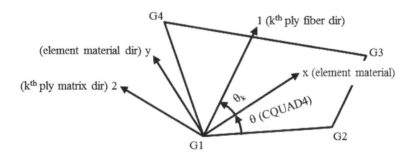

PLOAD2

Defines a pressure load on a shell element (CTRIA3 and CQUAD4)

(1)	(2)	(3)	(4)	(5)	(6)	(7)	(8)	(9)	(10)
PLOAD2	SID	P	EID1	EID2	EID3	EID4	EID5	EID6	

Field	Comments
SID	Load set identification number.

P   Pressure value. Positive pressure is in the direction of the element normal. (Real $\neq 0.0$)

EID#   Element identification numbers on which to apply the pressure.

PLOAD4

Defines a pressure load on a face of a solid element (CPENTA and CHEXA)

(1)	(2)	(3)	(4)	(5)	(6)	(7)	(8)	(9)	(10)
PLOAD4	SID	EID	P1	P2	P3	P4	G1	G3	
	CID	N1	N2	N3					

Field	Comments
SID	Load set identification number.

EID   Element identification numbers on which to apply the pressure.

P1/2/3/4   Pressure value at the corners of the face of the solid element. (Real $\neq 0.0$) (P1 is default for P2, P3, and P4)

G1   Grid point identification number of a grid point connected to a corner of the face of the solid element.

G3   Grid point identification number of a grid point connected to a corner of the face diagonally opposite to G1.

CID   Coordinate system identification number.

N1/2/3   Components of a vector in the defined coordinate system used to define the direction of the pressure. The magnitude of the pressures is set by P1 – P4 entries.

PLY

Defines a ply for composite ply-based shell modeling

(1)	(2)	(3)	(4)	(5)	(6)	(7)	(8)	(9)	(10)
PLY	ID	MID_k	t_k	θ_k	$SOUT_k$	$TMANUF_k$	DID_k		
	$ESID_1$	$ESID_2$	$ESID_3$	$ESID_4$	$ESID_5$	$ESID_6$	$ESID_7$	$ESID_8$	
	$ESID_9$	...							

Field	Comments
ID	Ply identification number.

MID_k Material identification number of the kth ply. Must refer to a MAT1, MAT2, or MAT8 card.

t_k Nominal thickness of the kth ply.

θ_k Nominal fiber orientation angle, in degrees, of the kth ply relative to the x-axis of the element material coordinate system. See figure below for θ conventions. (Default = 0.0)

$SOUT_k$ Stress, strain, and failure output option of the kth ply. Ply stress, strain, and failure output is given at the middle of each ply. In addition, OUTPUT CSTRESS and/or OUTPUT CSTRAIN cards must be defined in the I/O section to get output for the ply. (Default = NO)

             NO – do not output stress, strain, or failure for the kth ply.
             YES – output stress, strain, and failure for the kth ply

$TMANUF_k$ Actual manufactured ply thickness of the kth ply. This parameter is utilized in composite size optimization to automatically create discrete design variables such that the thickness of the ply bundle is equal to an integer multiple of TMANUF.

DID_k DRAPE data table identification number of the kth ply. A drape data table is used to define draping data for a ply. A drape data table defines a ply's actual fiber orientation angle and thickness by specifying variations from a ply's nominal fiber orientation angle and thickness at the centroid of each element that makes up the shape of a ply.

$ESID_i$ Element set identification numbers that define the elements that define shape of the ply. The superset of all elements defined by all referenced element set IDs define the shape of the ply.

Ply Fiber Orientation Angle

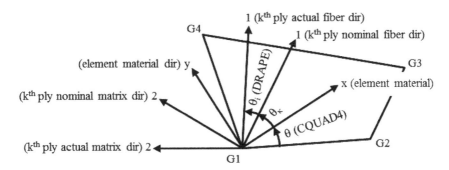

PSHELL

Defines the property definition of a homogeneous shell element.

(1)	(2)	(3)	(4)	(5)	(6)	(7)	(8)	(9)	(10)
PSHELL	PID	MID1	T	MID2	12I/T3	MID3	Ts/T	NSM	
	Z1	Z2	MID4	T0	ZOFFS				

Field   Comments

PID   Property identification number.

MID1   Material identification number for extension behavior of the plate. Must reference a MAT1, MAT2, or MAT8 card. This field must not be blank. If homogenizing by reference to a MAT2 card, see section 7.4 for calculation of the equivalent homogenized material matrix $[\overline{Q}_1]$.

T   Total thickness of the plate. Can by overridden by the Ti fields on the CQUAD4 and/or CTRIA3 cards.

MID2   Material identification number for bending behavior of the plate. If blank, then the plate has membrane behavior only. In addition, MID3 and MID4 fields must also be blank. If homogenizing by reference to a MAT2 card, see section 7.4 for calculation of the equivalent homogenized material matrix $[\overline{Q}_2]$.

12I/T3   Bending stiffness ratio of the plate. (Default = 1.0)

MID3   Material identification number for transverse shear behavior of the plate. If blank, then MID2 field is used to calculate the transverse shear behavior of the plate. If MID3 field is referenced by a MAT2 card, then Q_{33} field on the MAT2 card must be blank. If MID3 field is referenced by a MAT8 card, then G_{23} and G_{13} fields must not be blank. If homogenizing by reference to a MAT2 card, see section 7.4 for calculation of the equivalent homogenized material matrix $[\overline{Q}_3]$.

Ts/T   Transverse shear ratio of the plate. The transverse shear thickness divided by the total thickness of the homogenized plate. (Default = 0.8333)

NSM   Non-structural mass per unit area applied to the plate.

Z1   The first z-coordinate distance at which to calculate the stress and strain output for the plate, typically the bottom of the plate. (Default: $-T/2$)

Z2   The second z-coordinate distance at which to calculate the stress and strain output for the plate, typically the top of the plate. (Default: $T/2$)

MID4   Material identification number for extension-bending coupling behavior of the plate. Cannot reference the same material as the MID1 or MID2 fields. If homogenizing by reference to a MAT2 card, see section 7.4 for calculation of the equivalent homogenized material matrix $[\overline{Q_4}]$.

T0   Minimum homogeneous shell thickness. Valid for topology and free-size optimization with MAT1 card only. Can be overridden by the T0 field on the DSIZE card, THICK option. (Default = blank)

ZOFFS   Offset from the element grid point plane to the reference plane of the plate element. Can be overridden by the ZOFFS field on the CQUAD4 and/or CTRIA3 cards. See figures below for ZOFFS conventions.

Element offset

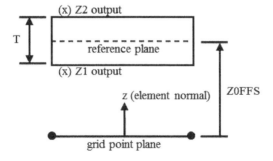

PSOLID

Defines the property definition of a homogeneous solid element.

(1)	(2)	(3)	(4)	(5)	(6)	(7)	(8)	(9)	(10)
PSOLID	PID	MID	CORDM						

Field	Comments
PID	Property identification number.

MID | Material identification number. Must reference a MAT1, MAT9, or MAT9ORT card.

CORDM | Material coordinate system identification number.

SET

Defines a set of nodes or elements.

(1)	(2)	(3)	(4)	(5)	(6)	(7)	(8)	(9)	(10)
SET	SID	TYPE	SUBTYPE						
	ID_1/X_1	ID_2/Y_1	ID_3/Z_1	ID_4/X_2	ID_5/Y_2	ID_6/Z_2	ID_7	ID_8	
	ID_9	...							

Alternate set card format for defining ID ranges.

(1)	(2)	(3)	(4)	(5)	(6)	(7)	(8)	(9)	(10)
SET	SID	TYPE	SUBTYPE						
	ID_1	THRU	ID_2	EXCEPT	ID_3	ID_4	ID_5	ID_6	
	ID_7	...	ENDTHRU						
	...								

Field	Comments
SID	Set identification number.

TYPE   Set type.

> **GRID** - Specifies that a set of node IDs are to follow.
> **ELEM** - Specifies that a set of element IDs are to follow.

SUBTYPE Set subtype.

> **LIST** - Specifies that a simple list of IDs the given set type are to follow. Valid for TYPE = GRID and ELEM.
> **BBOX** - Specifies that the coordinates of a bounding box are to follow. The coordinates (X_1,Y_1,Z_1) define one corner of the bounding box. The coordinates (X_2,Y_2,Z_2) define the opposite corner of the bounding box. All entities of TYPE within the bounding box are added to the set. Valid for TYPE = GRID and ELEM.
> **PROP** - Specifies that a list of property IDs are to follow. All elements that have a property ID equal to that of any of the listed property IDs are added to the element set. Valid for TYPE = ELEM.
> **MAT** - Specifies that a list of material IDs are to follow. All elements that have a property ID which references a material ID equal to that of any of the listed material IDs are added to the element set. Valid for TYPE = ELEM.

ID_i List of identification numbers for a given TYPE/SUBTYPE combination.

THRU   Keyword used to define ID ranges. All IDs between the preceding ID and the following ID are to be added to the set.

EXCEPT   Keyword used with ID range definitions to indicate that the following IDs are to be excluded from the current ID range definition.

ENDTHRU Keyword used after the EXCEPT keyword to define the end of an excluded ID list definition.

SPC

Defines a single point constraint and enforced displacement for static analysis.

(1)	(2)	(3)	(4)	(5)	(6)	(7)	(8)	(9)	(10)
SPC	SID	GID	C	D					

Field	Comments
SID	Single point constraint identification number.
GID	Grid identification number
C	Component numbers for the degrees of freedom to constraint. Up to six unique digits between 1 and 6. The component numbers refer to the coordinate system referenced by the CD field on the GRID card.
D	Enforced displacement for all component numbers defined.

STACK – Ply laminate definition

Defines the stacking sequence of a composite laminate for composite ply based modeling using the ply laminate definition of the STACK card.

(1)	(2)	(3)	(4)	(5)	(6)	(7)	(8)	(9)	(10)
STACK	ID	LAM	PLY ID$_1$	PLY ID$_2$	PLY ID$_3$	PLY ID$_4$	PLY ID$_5$	PLY ID$_6$	
	PLY ID$_7$	...	PLY ID$_n$						

Field	Comments
ID	Stack identification number.

LAM — Laminate stacking sequence option. If blank all plies must be specified. (Default = blank)

SYM - Only plies on the bottom half of the laminate need to be specified. This option is not valid for PCOMPG card.

MEM - All plies must be specified, however only [A] matrix terms are calculated. Therefore, the laminated plate exhibits extension behavior only. Any Z0 entry is ignored and set to the default value ($-T/2$).

BEND - All plies must be specified, however only [D] matrix terms are calculated. Therefore, the laminated plate exhibits bending behavior only. Any Z0 entry is ignored and set to the default value ($-T/2$).

SMEAR - All plies must be specified and SMEAR technology is utilized to calculate the ABD matrix of the laminate. Any Z0 entry is ignored and set to the default value ($-T/2$). See chapter 8 for details on SMEAR technology.

SMEARZ0 - All plies must be specified and SMEAR technology is utilized to calculate the ABD matrix of the laminate. The Z0 entry is considered in the calculation of the ABD matrix. Unlike SMEAR technology, SMEARZ0 will develop a B matrix due to the Z0 term. If Z0 is set to the default value ($-T/2$), then SMEAR and SMEARZ0 will produce the same ABD matrix. See chapter 8 for details on SMEAR technology.

SMCORE - All plies must be specified. The last ply specified must be the core layer. All other plies define the "top" and "bottom" face sheet laminates. Half of the total thickness of the laminate is placed on the "top" of the core. The other half of the laminate thickness is placed on the "bottom" of the core. SMEAR Core technology is utilized to calculate the ABD matrix of the laminate. Any Z0 entry is ignored and set to the default value ($-T/2$). See chapter 8 for details on SMEAR Core technology.

SYMEM - Only plies on the bottom half of laminate need to be specified, however only [A] matrix terms are calculated. Therefore, the laminated plate exhibits extension behavior only. Any Z0 entry is ignored and set to the default value ($-T/2$). This option is not valid for PCOMPG card.

SYBEND - Only plies on the bottom half of the laminate need to be specified, however only [D] matrix terms are calculated. Therefore, the laminated plate exhibits bending behavior only. Any Z0 entry is ignored and set to the default value ($-T/2$).This option is not valid for PCOMPG card.

SYSMEAR - Only plies on the bottom half of the laminate need to be specified and SMEAR technology is utilized to calculate the ABD matrix of the laminate. See chapter 8 for details on SMEAR technology.

$PLYID_k$ Ply identification number for the kth ply, defined in the order of the ply laminate stacking sequence as given in the figure below.

Laminate Stacking Sequence and Offset Definitions

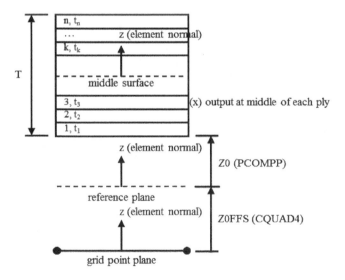

STACK - Interface laminate definition

Defines the stacking sequence of a composite laminate for composite ply based modeling using the interface laminate definition of the STACK card.

(1)	(2)	(3)	(4)	(5)	(6)	(7)	(8)	(9)	(10)
STACK	ID								
	SUB	...							
	...								
	INT	...							
	...								

Continuation lines to define sub-laminates.

(1)	(2)	(3)	(4)	(5)	(6)	(7)	(8)	(9)	(10)
	SUB	SID_i	$SNAME_i$	$SPLY$ ID_{i1}	$SPLY$ ID_{i2}	$SPLY$ ID_{i3}	$SPLY$ ID_{i4}	$SPLY$ ID_{i5}	
		$SPLY$ ID_{i6}	...	$SPLY$ ID_{1n}					

Continuation lines to define interface definitions.

(1)	(2)	(3)	(4)	(5)	(6)	(7)	(8)	(9)	(10)
	INT	$IPLYID_{i1}$	$IPLYID_{i2}$						

Field	Comments
ID	Stack identification number.

SUB   Keyword used to define the start of a sub-laminate definition data block. Multiple sub-laminate definitions can be defined on a single STACK card, each of which begins with the SUB keyword.

SID_i Sub-laminate identification number for the ith sub-laminate definition.

$SNAME_i$ Sub-laminate name for the ith sub-laminate definition.

$SPLYID_{ik}$ Sub-laminate ply identification number for the ith sub-laminate definition for the kth ply, defined in the order of the sub-laminate stacking sequence as defined in the figure below.

INT   Keyword used to define the start of an interface definition data block. Multiple interface definitions can be defined on a single STACK card, each of which begins with the INT keyword. Each interface definition defines exactly one interface laminate of a complete integrated structure.

$IPLYID_{i1}$

IPLYID$_{i2}$ Interface ply identification numbers defining the ith interface laminate definition. Interface plies can be either the 1st or nth ply of a sub-laminate definition and must come from different sub-laminates. IPLYID$_{i1}$ and IPLYID$_{i2}$ stack in the direction of the element normal at the interface between the two sub-laminates, which defines the directions the sub-laminates stack. The interface laminate stacking sequence follows directly as defined on the two sub-laminate stacking sequence definitions from their interface plies to the ply on the opposite side of the sub-laminate definition in their respective directions from the two interface plies as defined in the figure below.

<u>Sub-laminate Stacking Sequence Definition</u>

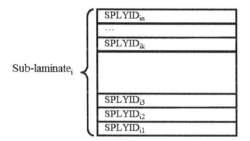

<u>Interface Laminate Stacking Sequence and Offset Definition</u>

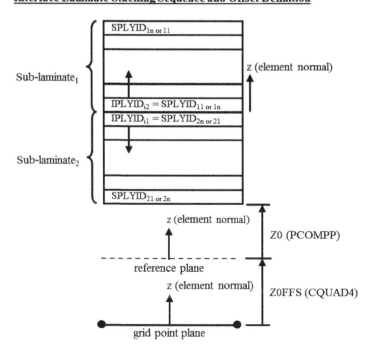

Appendix D
OptiStruct Optimization Bulk Data Reference

DCOMP

Defines composite size optimization design manufacturing constraints.

(1)	(2)	(3)	(4)	(5)	(6)	(7)	(8)	(9)	(10)
DCOMP	ID	ETYPE	EID_1	EID_2	EID_3	EID_4	EID_5	EID_6	
		EID_7	...						

Continuation to define total laminate thickness manufacturing constraints.

(1)	(2)	(3)	(4)	(5)	(6)	(7)	(8)	(9)	(10)
	LAMTHK	LTMIN	LTMAX	LTSET	LTEXC				

Continuation lines to define ply group thickness percentage manufacturing constraints.

(1)	(2)	(3)	(4)	(5)	(6)	(7)	(8)	(9)	(10)
	PLYPCT	PPGRP	PPMIN	PPMAX	PPOPT	PPSET	PPEXC		

Continuation lines to define ply group balancing manufacturing constraints.

(1)	(2)	(3)	(4)	(5)	(6)	(7)	(8)	(9)	(10)
	BALANCE	BGRP1	BGRP2		BOPT				

Continuation lines to define ply group constant thickness manufacturing constraints.

(1)	(2)	(3)	(4)	(5)	(6)	(7)	(8)	(9)	(10)
	CONST	CGRP	CTHICK		COPT				

Continuation lines to define ply group drop off manufacturing constraints.

(1)	(2)	(3)	(4)	(5)	(6)	(7)	(8)	(9)	(10)
	PLYDRP	PDGRP	PDTYP	PDMAX	PDOPT	PDSET	PDEXC		

Field   Comments
ID   Composite size optimization manufacturing constraint identification number.

ETYPE   Entity type.

           PCOMP – Specifies that the composite manufacturing constraints apply to a PCOMP zone-based laminate definition
           STACK – Specifies that the composite manufacturing constraints apply to a STACK ply-based laminate definition.

EID_i Entity identification number. Must be PCOMP identification numbers for ETYPE = PCOMP. Must be STACK identification numbers for ETYPE = STACK.

LAMTHK   Keyword used to define total laminate thickness manufacturing constraints. Multiple LAMTHK definitions can be defined on a single DCOMP card, each of which begins with the LAMTHK keyword.

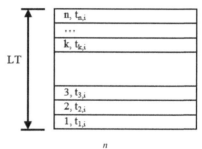

$$LT = \sum_{k=1}^{n} t_{k,i}$$

$$LTMIN < LT < LTMAX$$

LTMIN   Minimum laminate total thickness for the laminate total thickness manufacturing constraint. (LTMIN > 0.0)

LTMAX   Maximum laminate total thickness for the laminate total thickness manufacturing constraint. (LTMAX > LTMIN)

LTSET   Element set identification number defining the elements to which the laminate thickness manufacturing constraint applies. (Default = All Elements)

LTEXC   Ply exclusion option indicating plies that are to be excluded from the laminate thickness calculation for the laminate thickness manufacturing constraint. (Default = CORE)

NONE - No plies are excluded from the calculation.
CORE - The core layer within a SMCORE laminate definition (i.e. the last layer defined in the laminate definition) is excluded from the calculation. If the referenced PCOMP or STACK card is not defined as a SMCORE laminate, then there is no core layer defined.
CONST - Any ply defined with a CONST ply thickness manufacturing constraint is excluded from the calculation.
BOTH – Both the core layer within a SMCORE laminate definition and any ply defined with a CONST ply thickness manufacturing constraint are excluded from the calculation.

PLYPCT   Keyword used to define ply group percent thickness manufacturing constraints. Multiple PLYPCT definitions can be defined on a single DCOMP card, each of which begins with the PLYPCT keyword.

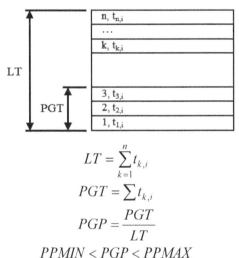

$$LT = \sum_{k=1}^{n} t_{k,i}$$
$$PGT = \sum t_{k,i}$$
$$PGP = \frac{PGT}{LT}$$
$$PPMIN < PGP < PPMAX$$

PPGRP   Ply group identification to which the ply group percent thickness manufacturing constraint applies. Ply groups can be identified by nominal fiber orientation angle, ply sets, or individual ply identification numbers depending on the PPOPT setting.

PPMIN   Minimum ply group percent thickness for the ply group percent thickness manufacturing constraint. (PPMIN > 0.0)

PPMAX   Maximum ply group percent thickness for the ply group percent thickness manufacturing constraint. (PPMAX > PTMIN)

PPOPT   Ply group identification option for the ply group percent thickness manufacturing constraint. (Default = BYANG)

BYANG - Specifies that PPGRP is defined as a real number representing a nominal fiber orientation angle. The ply group is defined as all the plies which have the given nominal fiber orientation angle.
BYSET - Specifies that PPGRP is defined as a ply set identification number. The ply group is the set of plies which are defined in the referenced ply set.
BYPLY – Specifies that PPGRP is defined as a single ply identification number. The ply group is the individual ply referenced by the ply identification number.

PPSET   Element set identification number defining the elements to which the ply group percent thickness manufacturing constraint applies. (Default = All Elements)

PPEXC   Ply exclusion option indicating plies that are to be excluded from the ply group percent thickness calculation for the ply group percent thickness manufacturing constraint. (Default = CORE)

NONE - No plies are excluded from the calculation.
CORE - The core layer within a SMCORE laminate definition (i.e. the last layer defined in the laminate definition) is excluded from the calculation. If the referenced PCOMP or STACK card is not defined as a SMCORE laminate, then there is no core layer defined.
CONST - Any ply defined with a CONST ply thickness manufacturing constraint is excluded from the calculation.
BOTH – Both the core layer within a SMCORE laminate definition and any ply defined with a CONST ply thickness manufacturing constraint are excluded from the calculation.

BALANCE Keyword used to define ply group balance manufacturing constraints. Multiple BALANCE definitions can be defined on a single DCOMP card, each of which begins with the BALANCE keyword.

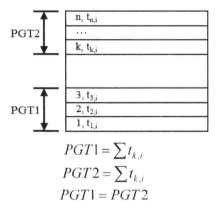

$$PGT1 = \sum t_{k,i}$$
$$PGT2 = \sum t_{k,i}$$
$$PGT1 = PGT2$$

BGRP1   Ply group #1 identification to which the ply group balance manufacturing constraint applies. Ply groups can be identified by nominal fiber orientation angle, ply sets, or individual ply identification numbers depending on the BOPT setting.

BGRP2   Ply group #2 identification to which the ply group balance manufacturing constraint applies. Ply groups can be identified by nominal fiber orientation angle, ply sets, or individual ply identification numbers depending on the BOPT setting.

BOPT   Ply group identification option for the ply group balance manufacturing constraint. (Default = BYANG)

BYANG - Specifies that BGRP1 and BGRP2 are defined as real numbers representing the nominal fiber orientation angles. The ply group is defined as all the plies which have the given nominal fiber orientation angle.

BYSET - Specifies that BGRP1 and BRRP2 are defined as ply set identification numbers. The ply group is the set of plies which are defined in the referenced ply set.

BYPLY – Specifies that BGRP1 and BGRP2 are defined as single ply identification numbers. The ply group is the individual ply referenced by the ply identification number.

CONST   Keyword used to define ply group constant thickness manufacturing constraints. Multiple CONST definitions can be defined on a single DCOMP card, each of which begins with the CONST keyword.

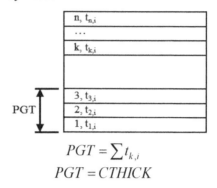

$$PGT = \sum t_{k,i}$$
$$PGT = CTHICK$$

CGRP   Ply group identification to which the ply group constant thickness manufacturing constraint applies. Ply groups can be identified by nominal fiber orientation angle, ply sets, or individual ply identification numbers depending on the COPT setting.

CTHICK   Constant ply group thickness for the ply group constant thickness manufacturing constraint. (CTHICK > 0.0)

COPT   Ply group identification option for the ply group constant thickness manufacturing constraint. (Default = BYANG)

          BYANG - Specifies that CGRP is defined as a real number representing the nominal fiber orientation angle. The ply group is defined as all the plies which have the given nominal fiber orientation angle.
          BYSET - Specifies that CGRP is defined as a ply set identification number. The ply group is the set of plies which are defined in the referenced ply set.
          BYPLY – Specifies that CGRP is defined as a single ply identification number. The ply group is the individual ply referenced by the ply identification number.

PLYDRP   Keyword used to define ply group drop off manufacturing constraints. Multiple PLYDRP definitions can be defined on a single DCOMP card, each of which begins with the PLYDRP keyword.

PDGRP   Ply group identification to which the ply group drop off manufacturing constraint applies. Ply groups can be identified by nominal fiber orientation angle, ply sets, or individual ply identification numbers depending on the PDOPT setting.

PDTYPE   Specifies type of ply group drop of manufacturing constraint to apply. (Default = TOTDRP)

          TOTDRP uses the total laminate drop method to calculate the ply drop manufacturing constraint.

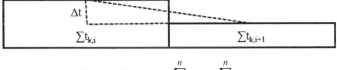

$$PDMAX = \Delta t = \sum_{k=1}^{n} t_{k,i} - \sum_{k=1}^{n} t_{k,i+1}$$

PDMAX   Maximum ply group drop off based on the PDTYPE setting. (PPMAX > 0)

PDOPT   Ply group identification option for the ply group drop off manufacturing constraint. (Default = BYANG)

          BYANG - Specifies that PPGRP is defined as a real number representing a nominal fiber orientation angle. The ply group is defined as all the plies which have the given nominal fiber orientation angle.
          BYSET - Specifies that PPGRP is defined as a ply set identification number. The ply group is the set of plies which are defined in the referenced ply set.
          BYPLY – Specifies that PPGRP is defined as a single ply identification number. The ply group is the individual ply referenced by the ply identification number.

PDSET   Element set identification number defining the elements to which the ply group drop off manufacturing constraint applies. (Default = All Elements)

PDEXC   Ply exclusion option indicating plies that are to be excluded from the ply group drop off calculation for the ply group drop off manufacturing constraint. (Default = CORE)

NONE - No plies are excluded from the calculation.

CORE - The core layer within a SMCORE laminate definition (i.e. the last layer defined in the laminate definition) is excluded from the calculation. If the referenced PCOMP or STACK card is not defined as a SMCORE laminate, then there is no core layer defined.

CONST - Any ply defined with a CONST ply thickness manufacturing constraint is excluded from the calculation.

BOTH – Both the core layer within a SMCORE laminate definition and any ply defined with a CONST ply thickness manufacturing constraint are excluded from the calculation.

DCONADD

Defines an optimization design constraint as a combination of DCONSTR design constraint definitions.

(1)	(2)	(3)	(4)	(5)	(6)	(7)	(8)	(9)	(10)
DCONADD	DCID	DC_1	DC_2	DC_3	DC_4	DC_5	DC_6	DC_7	
		DC_8	...						

Field | Comments
DCID | Design constraint identification number.

DC_i | DCONSTR identification numbers used to create a new design constraint as the combination of referenced DCONSTR identification numbers.

DCONSTR

Defines an optimization design constraint.

(1)	(2)	(3)	(4)	(5)	(6)	(7)	(8)	(9)	(10)
DCONSTR	DCID	RID	LBOUND	UBOUND	LFREQ	UFREQ			

Field   Comments

DCID   Design constraint identification number.

RID   Design response identification number for which to apply the design constraint.

LBOUND   Design constraint lower bound value for the referenced design response.

UBOUND   Design constraint upper bound value for the referenced design response.

LFREQ   Design constraint lower bound frequency value. This value only applies to frequency design responses related to frequency response subcases. The design constraints bounds, LBOUND and UBOUND, are applied only if the loading frequency falls between LFREQ and UFREQ.

UFREQ   Design constraint upper bound frequency value. This value only applies to frequency design responses related to frequency response subcases. The design constraints bounds, LBOUND and UBOUND, are applied only if the loading frequency falls between LFREQ and UFREQ.

DDVAL

Defines a discrete design value list for an optimization design variable.

(1)	(2)	(3)	(4)	(5)	(6)	(7)	(8)	(9)	(10)
DDVAL	ID	DVAL$_1$	DVAL$_2$	DVAL$_3$	DVAL$_4$	DVAL$_5$	DVAL$_6$	DVAL$_7$	
	DVAL$_8$	...							

Alternate form of the DDVAL card:

(1)	(2)	(3)	(4)	(5)	(6)	(7)	(8)	(9)	(10)
DDVAL	ID	DVAL$_1$	THRU	DVAL$_2$	BY	INC			
	DVAL$_1$	THRU	DVAL$_2$	BY	INC				
	...								

Field	Comments
ID	Discrete design value list identification number.
DVAL$_i$	Discrete design values. Can be listed in any order.
THRU	Keyword used in alternate discrete design value list definition.
BY	Keyword used in alternate discrete design value list definition.
INC	Discrete design value increment. The list of discrete design values generated by the alternate format is $DVAL_1 + (n)(INC)$, where n = 0, 1, 2, ...n. The last discrete design value is always $DVAL_2$ even if the range is not evenly divisible by INC.

DESVAR

Defines an optimization design variable.

(1)	(2)	(3)	(4)	(5)	(6)	(7)	(8)	(9)	(10)
DESVAR	ID	LABEL	XINIT	XLB	XUB	DELXV	DDVAL		

Field	Comments
ID	Design variable identification number.

LABEL   User-defined label for the design variable.

XINIT   Design variable initial value.

XLB   Design variable lower bound value.

XUB   Design variable upper bound value.

DELXV   Design variable initial move limit.
For size optimization the value is a fraction of the design variable. (Default = Value of DOPTPRM, DELSIZ parameter)
For shape optimization the value is a fraction of the range (XUB − XLB). (Default − Value of DOPTRM, DELSHP parameter)

DDVAL   Identification number of a discrete design value list (DDVAL) defining a set of permissible discrete design values.

DRESP1

Defines an optimization design response.

(1)	(2)	(3)	(4)	(5)	(6)	(7)	(8)	(9)	(10)
DRESP1	ID	LABEL	RTYPE	PTYPE	REGION	ATTA	ATTB	ATT_1	
	ATT_2	...							
	EXCL	EID_1	EID_2	EID_3	EID_4	EID_5	EID_6	EID_7	
		EID_8	...						

Field | Comments

ID   Design response identification number.

LABEL   User-defined label for the design response.

RTYPE   Design response type. See the design responses table below.

PTYPE   Design response property type where required. See the design responses table below.

REGION   Design response region identifier. (Integer > 0, Default = blank)

ATTA   Design response attribute A where required. See the design responses table below.

ATTB   Design response attribute B where required. See the design responses table below.

ATT_i Identification numbers of the appropriate PTYPE. See the design responses table below.

EXCL   Keyword indicating that element identification numbers follow that are to be excluded from the design response.

EID_i Element identification numbers from which design responses will not be generated.

Design Responses Table

Response	RTYPE	PTYPE	ATTA	ATTB	ATT$_i$
Mass	MASS	Property Card or MAT	-	COMB SUM	PID MID Blank = All
Mass Fraction	MASSFRAC	Property Card or MAT	-	COMB SUM	PID MID Blank = All
Volume	VOLUME	Property Card or MAT	-	COMB SUM	PID MID Blank = All
Volume Fraction	VOLFRAC	Property Card or MAT	-	COMB SUM	PID MID Blank = All
Center of Gravity	COG	Property Card or MAT	Center of Gravity Item Code	COMB	PID MID Blank = All
Moment of Inertia	INERTIA	Property Card or MAT	Moment of Inertia Item Code	COMB	PID MID Blank = All
Compliance	COMP	Property Card or MAT	-	COMB SUM	PID MID Blank = All
Weighted Compliance	WCOMP	Property Card or MAT	-	-	PID MID Blank = All
Displacement	DISP	-	Displacement Component Item Code	-	GID
Normal Mode Frequency	FREQ	-	Normal Mode Number	-	-
Buckling Mode Eigenvalue	LAMA	-	Buckling Mode Number	-	-
Homogeneous Stress (Z1, Z2)	STRESS	Property Card or ELEM	Homogeneous Stress Item Code	-	PID EID Blank = All
Homogeneous Strain (Z1, Z2)	STRAIN	Property Card or ELEM	Homogeneous Strain Item Code	-	PID EID Blank = All
Composite Ply Stress (mid of ply)	CSTRESS	PCOMPG PLY ELEM	Composite Ply Stress Item Code	ALL G# (global ply #)	PID EID PLYID Blank = All
Composite Ply Strain (mid of ply)	CSTRAIN	PCOMPG PLY ELEM	Composite Ply Strain Item Code	ALL G# (global ply #)	PID EID PLYID Blank = All
Composite Ply Failure (mid of ply)	CFAILURE	PCOMPG PLY ELEM	Composite Ply Failure Item Code	ALL G# (global ply #)	PID EID PLYID Blank = All
Static Force	FORCE	Property Card or MAT	Static Force Item Code	-	PID EID Blank = All
SPC Forces	SPCFORCE		Component DOF (1-6)		GID
GPF Balance	GPFORCE	GID	Component DOF (1-6)		EID

Center of Gravity Item Codes

Component	Code
x-coordinate	X
y-coordinate	Y
z-coordinate	Z

Moment of Inertia Item Codes

Component	Code
Ixx	XX
Iyy	YY
Izz	ZZ
Ixy	XY
Iyz	YZ
Ixz	XZ

Homogeneous Stress or Strain Item Codes

Element	Component	ASCII Code
All Solid Elements	$\sigma_{vm} / \varepsilon_{vm}$	SVM
	σ_1 / ε_1	SMAP
	σ_2 / ε_2	SMDP
	σ_3 / ε_3	SMIP
	σ_x / ε_x	SXX
	σ_y / ε_y	SYY
	σ_z / ε_z	SZZ
	τ_{xy} / γ_{xy}	SXY
	τ_{yz} / γ_{yz}	SYZ
	τ_{xz} / γ_{xz}	SXZ
All Shell Elements	$\sigma_{vm} / \varepsilon_{vm}$	SVMB
(Both Sides)	σ_1 / ε_1	SMPB
	σ_3 / ε_3	SMIPB
	σ_x / ε_x	SXB
	σ_y / ε_y	SYB
	τ_{xy} / γ_{xy}	SXYB
(Z1)	$\sigma_{vm} / \varepsilon_{vm}$	SVM1
	σ_1 / ε_1	SMP1
	σ_3 / ε_3	SMIP1
	σ_x / ε_x	SX1
	σ_y / ε_y	SY1
	τ_{xy} / γ_{xy}	SXY1
(Z2)	$\sigma_{vm} / \varepsilon_{vm}$	SVM2
	σ_1 / ε_1	SMP2
	σ_3 / ε_3	SMIP2
	σ_x / ε_x	SX2
	σ_y / ε_y	SY2
	τ_{xy} / γ_{xy}	SXY2

Composite Ply Stress or Strain Item Codes

Type	Component	Code
(Total Stress/Strain)	σ_1 / ε_1	S1
	σ_2 / ε_2	S2
	τ_{12} / γ_{12}	S12
	τ_{23} / γ_{23}	S2Z
	τ_{13} / γ_{13}	S1Z
	σ_1 / ε_1 (principal)	SMAP
	σ_3 / ε_3 (principal)	SMIP
(Mechanical Strain)	$\varepsilon_{m,1}$	MS1
	$\varepsilon_{m,2}$	MS2
	$\gamma_{m,12}$	MS12
	$\varepsilon_{m,1}$ (principal)	MSMAP
	$\varepsilon_{m,3}$ (principal)	MSMIP
(Thermal Strain)	$\varepsilon_{t,1}$	TS1
	$\varepsilon_{t,2}$	TS2
	$\gamma_{t,12}$	TS12
	$\varepsilon_{t,1}$ (principal)	TSMAP
	$\varepsilon_{t,3}$ (principal)	TSMIP

Composite Ply Failure Item Codes

Theory	Code
Maximum Strain	STRN
Tsai-Hill	HILL
Tsai-Wu	TSAI

Static Force Item Codes

Element	Component	Code
All Shell Elements	N_x	NX
	N_y	NY
	N_{xy}	NXY
	M_x	MX
	M_y	MY
	M_{xy}	MXY
	Q_x	SXZ
	Q_y	SYZ

DSHUFFLE

Defines shuffling optimization design variables and design manufacturing constraints.

(1)	(2)	(3)	(4)	(5)	(6)	(7)	(8)	(9)	(10)
DSHUFFLE	ID	PTYPE	PID_1	PID_2	PID_3	PID_4	PID_5	PID_6	
		PID_7	...						

Continuation lines to define a maximum successive layers constraint.

(1)	(2)	(3)	(4)	(5)	(6)	(7)	(8)	(9)	(10)
	MAXSUCC	MANGLE	MSUCC	VSUCC					

Continuation line to define a pairing constraint.

(1)	(2)	(3)	(4)	(5)	(6)	(7)	(8)	(9)	(10)
	PAIR	PANGLE1	PANGLE2	POPT					

Continuation line to define a core layer stacking sequence constraint.

(1)	(2)	(3)	(4)	(5)	(6)	(7)	(8)	(9)	(10)
	CORE	CREP	CANG1	CANG2	CANG3	CANG4	CANG5	CANG6	
			CANG7	...					

Continuation line to define a cover layer stacking sequence constraint.

(1)	(2)	(3)	(4)	(5)	(6)	(7)	(8)	(9)	(10)
	COVER	VREP	VANG1	VANG2	VANG3	VANG4	VANG5	VANG6	
			VANG7	...					

Field	Comments
ID	Free-size optimization design variable and design manufacturing constraint identification number
PTYPE	Property type on which to apply the free-size design variables and design manufacturing constraints.

PCOMP - Defines PCOMP identification numbers follow.
PCOMPG - Defines PCOMPG identification numbers follow.
STACK – Defines STACK identification numbers follow.

PID_i	List of property identification numbers of PTYPE on which to apply the shuffle optimization design variables and design manufacturing constraints.

MAXSUCC Keyword used to define shuffling optimization maximum number of successive plies for a given angle constraint.

MANGLE Ply orientation, in degrees, to which the MAXSUCC constraint is applies.

MSUCC   Maximum number of successive plies for the angle defined by MANGLE. (Integer > 0)

VSUCC   Allowable percentage violation for the MAXSUCC constraint. 0.0 indicates that this constraint cannot be violated (Default = 0.0)

PAIR   Keyword used to define shuffling optimization pairing constraint.

PANGLE1 First ply orientation, in degrees, to which the PAIR constraint is applied. (only 45.0 allowed at this time)

PANGLE2 Second ply orientation, in degrees, to which the PAIR constraint is applies. (only -45.0 allowed at this time)

POPT   Pairing constraint option. SAME indicates that the stacking sequence should remain the same for consecutive pairs. REVERSE indicates that the stacking sequence should be reversed for alternate pairs.

CORE   Keyword used to define shuffling optimization core layer stacking sequence constraint. The core layer is defined by the plies around the middle surface of the laminate.

CREP   Number of times the core layer stacking sequence should be repeated (Integer > 0, Default = 1)

CANG#   Ply orientations, in degrees, defining the core layer stacking sequence.

COVER   Keyword used to define shuffling optimization cover layer stacking sequence constraint. The cover layer is defined by the plies at the top/bottom surface of the laminate.

VREP   Number of times the cover layer stacking sequence should be repeated (Integer > 0, Default = 1)

VANG#   Ply orientations, in degrees, defining the cover layer stacking sequence

DSIZE

Defines free-size optimization design variables and design manufacturing constraints.

(1)	(2)	(3)	(4)	(5)	(6)	(7)	(8)	(9)	(10)
DSIZE	ID	PTYPE	PID_1	PID_2	PID_3	PID_4	PID_5	PID_6	
		PID_7	...						

Continuation lines to define zone-based free-size optimization groups.

(1)	(2)	(3)	(4)	(5)	(6)	(7)	(8)	(9)	(10)
	GROUP		EG_1	EG_2	EG_3	EG_4	EG_5	EG_6	
		EG_7	...						

Continuation line to define a homogeneous shell thickness constraint.

(1)	(2)	(3)	(4)	(5)	(6)	(7)	(8)	(9)	(10)
	THICK	T0	T1						

Continuation line to define a homogeneous shell von Mises stress constraint.

(1)	(2)	(3)	(4)	(5)	(6)	(7)	(8)	(9)	(10)
	STRESS	UBOUND							

Continuation line to define a member size manufacturing constraint.

(1)	(2)	(3)	(4)	(5)	(6)	(7)	(8)	(9)	(10)
	MEMBSIZ	MINDIM							

Continuation lines to define total laminate thickness manufacturing constraints.

(1)	(2)	(3)	(4)	(5)	(6)	(7)	(8)	(9)	(10)
	COMP	LAMTHK	LTMIN	LTMAX	LTSET	LTEXC			

Continuation lines to define ply group thickness percentage manufacturing constraints.

(1)	(2)	(3)	(4)	(5)	(6)	(7)	(8)	(9)	(10)
	COMP	PLYPCT	PPGRP	PPMIN	PPMAX	PPOPT	PPSET	PPEXC	

Continuation lines to define ply group balancing manufacturing constraints.

(1)	(2)	(3)	(4)	(5)	(6)	(7)	(8)	(9)	(10)
	COMP	BALANCE	BGRP1	BGRP2		BOPT			

Continuation lines to define ply group constant thickness manufacturing constraints.

(1)	(2)	(3)	(4)	(5)	(6)	(7)	(8)	(9)	(10)
	COMP	CONST	CGRP	CTHICK		COPT			

Continuation lines to define ply group drop off manufacturing constraints.

(1)	(2)	(3)	(4)	(5)	(6)	(7)	(8)	(9)	10
	COMP	PLYDRP	PDGRP	PDTYPE	PDMAX	PDOPT	PDSET	PDEXC	
			PDDEF	PDX	PDY	PDZ			

Field	Comments
ID	Free-size optimization design variable and design manufacturing constraint identification number
PTYPE	Property type on which to apply the free-size design variables and design manufacturing constraints.

PCOMP - Defines PCOMP identification numbers follow.
PCOMPG - Defines PCOMPG identification numbers follow.
PSHELL - Defines PSHELL identification numbers follow.
STACK – Defines STACK identification numbers follow.

PID_i List of property identification numbers of PTYPE on which to apply the free-size optimization design variables and design manufacturing constraints. For PTYPE = PCOMP, PCOMPG, or STACK; design variables are automatically created for the thickness of each ply for each element referenced by the property identification numbers. For PTYPE = PSHELL; design variables are automatically created for the thickness of the homogeneous shell for each element referenced by the property identification numbers.

GROUP   Keyword used to define free-sizing optimization zone groups. Free-size optimization zone groups are defined by element sets. For zone groups, design variables are automatically created for the thickness of each ply for each zone group. Therefore, the thickness of the plies within a zone group will be uniform and no ply drops or additions will exist within the zone group. Effectively, a zone group removes the element-by-element nature of the free-size optimization design variables and considers all the elements within the zone group as the same.

EG$_i$ Element set identification numbers defining the element within the zone group.

THICK   Keyword used to define a homogeneous thickness constraint. This keyword is valid for PTYPE = PSHELL only. The THICK keyword can be defined only once on a DSIZE card.

T0   Minimum homogeneous shell thickness for the homogeneous thickness constraint. Overrides the T0 field on the PSHELL card. If no value is entered, then the T0 field on the PSHELL card is used. If no value is entered on the PSHELL card, then T0 = 0.0 is assumed. (Default = blank, T0 > 0.0)

T1   Maximum homogeneous shell thickness for the homogeneous thickness constraint. If no value is entered, then T field on the PSHELL card is used. (Default = blank, T1 > T0)

STRESS   Keyword used to define a homogeneous von Mises stress constraint. This keyword is valid for PTYPE = PSHELL only. The STRESS keyword can be defined only once on a DSIZE card.

UBOUND   Upper bound value for the homogeneous von Mises stress constraint. The von Mises stress cannot exceed this value. (UBOUND > 0.0)

MEMBSIZ   Keyword used to define a member size manufacturing constraint. The MEMBSIZ keyword can be defined only once on a DSIZE card.

MINDIM   Minimum dimension of formed members for the member size manufacturing constraint. Used to prevent the formation of small members. (Default = blank – no minimum size control, MINDIM > 0.0).

COMP   Keyword used to indicate the start of a composite manufacturing constraint definition. The COMP keyword is followed by the specific composite manufacturing constraint keyword being defined; LAMTHK, PLYTHK, PLYPCT, BALANCE, CONST, or PLYDRP.

LAMTHK   Keyword used to define total laminate thickness manufacturing constraints. Multiple LAMTHK definitions can be defined on a single DCOMP card, each of which begins with the LAMTHK keyword.

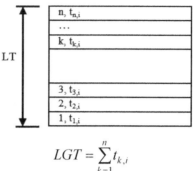

$$LGT = \sum_{k=1}^{n} t_{k,i}$$

$$LTMIN < LGT < LTMAX$$

LTMIN   Minimum laminate total thickness for the laminate total thickness manufacturing constraint. (LTMIN > 0.0)

LTMAX   Maximum laminate total thickness for the laminate total thickness manufacturing constraint. (LTMAX > LTMIN)

LTSET   Element set identification number defining the elements to which the laminate thickness manufacturing constraint applies. (Default = All Elements)

LTEXC   Ply exclusion option indicating plies that are to be excluded from the laminate thickness calculation for the laminate thickness manufacturing constraint. (Default = CORE)

NONE - No plies are excluded from the calculation.
CORE - The core layer within a SMCORE laminate definition (i.e. the last layer defined in the laminate definition) is excluded from the calculation. If the referenced PCOMP or STACK card is not defined as a SMCORE laminate, then there is no core layer defined.
CONST - Any ply defined with a CONST ply thickness manufacturing constraint is excluded from the calculation.
BOTH – Both the core layer within a SMCORE laminate definition and any ply defined with a CONST ply thickness manufacturing constraint are excluded from the calculation.

PLYPCT   Keyword used to define ply group percent thickness manufacturing constraints. Multiple PLYPCT definitions can be defined on a single DCOMP card, each of which begins with the PLYPCT keyword.

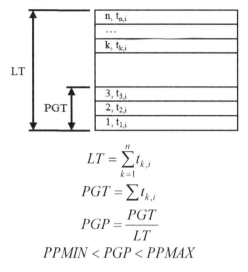

$$LT = \sum_{k=1}^{n} t_{k,i}$$

$$PGT = \sum t_{k,i}$$

$$PGP = \frac{PGT}{LT}$$

$$PPMIN < PGP < PPMAX$$

PPGRP   Ply group identification to which the ply group percent thickness manufacturing constraint applies. Ply groups can be identified by nominal fiber orientation angle, ply sets, or individual ply identification numbers depending on the PPOPT setting.

PPMIN   Minimum ply group percent thickness for the ply group percent thickness manufacturing constraint. (PPMIN > 0.0)

PPMAX   Maximum ply group percent thickness for the ply group percent thickness manufacturing constraint. (PPMAX > PTMIN)

PPOPT   Ply group identification option for the ply group percent thickness manufacturing constraint. (Default = BYANG)

BYANG - Specifies that PPGRP is defined as a real number representing a nominal fiber orientation angle. The ply group is defined as all the plies which have the given nominal fiber orientation angle.
BYSET - Specifies that PPGRP is defined as a ply set identification number. The ply group is the set of plies which are defined in the referenced ply set.

BYPLY – Specifies that PPGRP is defined as a single ply identification number. The ply group is the individual ply referenced by the ply identification number.

PPSET   Element set identification number defining the elements to which the ply group percent thickness manufacturing constraint applies. (Default = All Elements)

PPEXC   Ply exclusion option indicating plies that are to be excluded from the ply group percent thickness calculation for the ply group percent thickness manufacturing constraint. (Default = CORE)

NONE - No plies are excluded from the calculation.
CORE - The core layer within a SMCORE laminate definition (i.e. the last layer defined in the laminate definition) is excluded from the calculation. If the referenced PCOMP or STACK card is not defined as a SMCORE laminate, then there is no core layer defined.
CONST - Any ply defined with a CONST ply thickness manufacturing constraint is excluded from the calculation.
BOTH – Both the core layer within a SMCORE laminate definition and any ply defined with a CONST ply thickness manufacturing constraint are excluded from the calculation.

BALANCE Keyword used to define ply group balance manufacturing constraints. Multiple BALANCE definitions can be defined on a single DCOMP card, each of which begins with the BALANCE keyword.

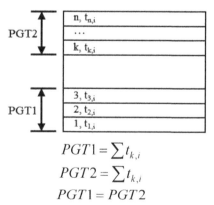

$$PGT1 = \sum t_{k,i}$$
$$PGT2 = \sum t_{k,i}$$
$$PGT1 = PGT2$$

BGRP1   Ply group #1 identification to which the ply group balance manufacturing constraint applies. Ply groups can be identified by nominal fiber orientation angle, ply sets, or individual ply identification numbers depending on the BOPT setting.

BGRP2   Ply group #2 identification to which the ply group balance manufacturing constraint applies. Ply groups can be identified by nominal fiber orientation angle, ply sets, or individual ply identification numbers depending on the BOPT setting.

BOPT   Ply group identification option for the ply group balance manufacturing constraint. (Default = BYANG)

BYANG - Specifies that BGRP1 and BGRP2 are defined as real numbers representing the nominal fiber orientation angles. The ply group is defined as all the plies which have the given nominal fiber orientation angle.
BYSET - Specifies that BGRP1 and BRRP2 are defined as ply set identification numbers. The ply group is the set of plies which are defined in the referenced ply set.
BYPLY – Specifies that BGRP1 and BGRP2 are defined as single ply identification numbers. The ply group is the individual ply referenced by the ply identification number.

CONST   Keyword used to define ply group constant thickness manufacturing constraints. Multiple CONST definitions can be defined on a single DCOMP card, each of which begins with the CONST keyword.

$$PGT = \sum t_{k,i}$$
$$PGT = CTHICK$$

CGRP   Ply group identification to which the ply group constant thickness manufacturing constraint applies. Ply groups can be identified by nominal fiber orientation angle, ply sets, or individual ply identification numbers depending on the COPT setting.

CTHICK   Constant ply group thickness for the ply group constant thickness manufacturing constraint. (CTHICK > 0.0)

COPT   Ply group identification option for the ply group constant thickness manufacturing constraint. (Default = BYANG)

           BYANG - Specifies that CGRP is defined as a real number representing the nominal fiber orientation angle. The ply group is defined as all the plies which have the given nominal fiber orientation angle.
           BYSET - Specifies that CGRP is defined as a ply set identification number. The ply group is the set of plies which are defined in the referenced ply set.
           BYPLY – Specifies that CGRP is defined as a single ply identification number. The ply group is the individual ply referenced by the ply identification number.

PLYDRP   Keyword used to define ply group drop off manufacturing constraints. Multiple PLYDRP definitions can be defined on a single DCOMP card, each of which begins with the PLYDRP keyword.

PDGRP   Ply group identification to which the ply group drop off manufacturing constraint applies. Ply groups can be identified by nominal fiber orientation angle, ply sets, or individual ply identification numbers depending on the PDOPT setting.

PDTYPE   Specifies type of ply group drop of manufacturing constraint to apply. (Default = PLYSLP)

           PLYSLP uses the ply slope method to calculating ply drop manufacturing constraint.

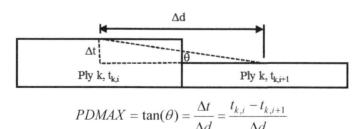

$$PDMAX = \tan(\theta) = \frac{\Delta t}{\Delta d} = \frac{t_{k,i} - t_{k,i+1}}{\Delta d}$$

PLYDRP uses the ply drop method to calculating ply drop manufacturing constraint

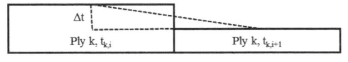

$$PDMAX = \Delta t = t_{k,i} - t_{k,i+1}$$

TOTSLP uses the total laminate slope method to calculating ply drop manufacturing constraint. Same as PLYSLP but considering ALL the plies as a total thickness, not just the kth ply, in PDGRP.

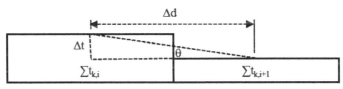

$$PDMAX = \tan(\theta) = \frac{\Delta t}{\Delta d} = \frac{\sum_{k=1}^{n} t_{k,i} - \sum_{k=1}^{n} t_{k,i+1}}{\Delta d}$$

TOTDRP uses the total laminate drop method to calculating ply drop manufacturing constraint. Same as PLYDRP but considering ALL the plies as a total thickness, not just the kth ply, in PDGRP.

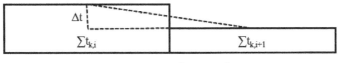

$$PDMAX = \Delta t = \sum_{k=1}^{n} t_{k,i} - \sum_{k=1}^{n} t_{k,i+1}$$

PDMAX   Maximum ply group drop off based on the PDTYPE setting. (PPMAX > 0)

PDOPT   Ply group identification option for the ply group drop off manufacturing constraint. (Default = BYANG)

BYANG - Specifies that PPGRP is defined as a real number representing a nominal fiber orientation angle. The ply group is defined as all the plies which have the given nominal fiber orientation angle.

BYSET - Specifies that PPGRP is defined as a ply set identification number. The ply group is the set of plies which are defined in the referenced ply set.

BYPLY – Specifies that PPGRP is defined as a single ply identification number. The ply group is the individual ply referenced by the ply identification number.

PDSET   Element set identification number defining the elements to which the ply group drop off manufacturing constraint applies. (Default = All Elements)

PDEXC   Ply exclusion option indicating plies that are to be excluded from the ply group drop off calculation for the ply group drop off manufacturing constraint. (Default = CORE)

          NONE - No plies are excluded from the calculation.

          CORE - The core layer within a SMCORE laminate definition (i.e. the last layer defined in the laminate definition) is excluded from the calculation. If the referenced PCOMP or STACK card is not defined as a SMCORE laminate, then there is no core layer defined.

          CONST - Any ply defined with a CONST ply thickness manufacturing constraint is excluded from the calculation.

          BOTH – Both the core layer within a SMCORE laminate definition and any ply defined with a CONST ply thickness manufacturing constraint are excluded from the calculation.

PDDEF   Optional definition to fine-tune the ply group drop off manufacturing constraint by requesting directional drop off. DIRECT is currently the only option available.

PDX/Y/Z   Used to specify the drop off direction when DIRECT is used in the PDDEF field. Defines the components of a direction vector, in the global coordinate system, in which the drop off constraint is to be applied. For example, if drop off control is required in the x-axis direction, then 1,0,0 should be entered for PDX, PDY, PDZ respectively.

DVPREL1

Defines property values, at each ith iteration of a size optimization, as a linear combination of design variables.

$$P_i = C_0 + \sum (COEF_i)(DVID_i)$$

(1)	(2)	(3)	(4)	(5)	(6)	(7)	(8)	(9)	(10)
DVPREL1	ID	TYPE	PID	PNAME or FID			C_0		
	$DVID_1$	$COEF_1$	$DVID_2$	$COEF_2$	$DVID_3$	$COEF_3$	$DVID_4$	$COEF_4$	
	$DVID_5$	$COEF_5$	...						

Field	Comments
ID	Design variable property relationship identification number.

TYPE — Property type.

PID — Property identification number of TYPE.

PNAME — Property field variable name. (i.e. T for the thickness of a ply on the PLY card)

FID — Property field identification number. The first row has field identification numbers 1 – 10, the second row has field identification numbers 11 – 20, and so on. (i.e. 4 for the thickness of a ply on the PLY card)

C_0 — Constant in the linear combination equation. (Default = 0.0)

$DVID_i$ — Design variable identification numbers defining the design variable to link in the linear combination equation.

$COEF_i$ — Design variable coefficients in the linear combination equation. (Default = 1.0)

Appendix E
OptiStruct Composite Analysis Models

OptiStruct Composite Zone-Based Shell Model

```
TITLE = 10" x 10" Plate
SUBTITLE = Zone-Based Shell Model
$Output File Formats
OUTPUT,H3D
OUTPUT,OPTI
$Global Output Requests
DISPLACEMENT = ALL
CSTRAIN(MECH,THER) = ALL
CSTRESS = ALL
STRESS = NO
$Global Subcase Control
TEMPERATURE(INITIAL) = 4
$Subcase Definitions
SUBCASE 1
  LABEL Nx_step
  ANALYSIS STATICS
  SPC = 1
  LOAD = 2
SUBCASE 2
  LABEL Nxt_step
  ANALYSIS STATICS
  SPC = 1
  TEMPERATURE(LOAD) = 3
SUBCASE 3
  LABEL Nx_Nxt_step
  ANALYSIS STATICS
  SPC = 1
  LOAD = 2
  TEMPERATURE(LOAD) = 3
BEGIN BULK
$--1---|---2---|---3---|---4---|---5---|---6---|---7---|---8---|---9---|
GRID           1           0.0      0.0      0.0
GRID           2           5.0     -5.0      0.0
GRID           3           5.0      5.0      0.0
GRID           4          -5.0      5.0      0.0
GRID           5          -5.0     -5.0      0.0
GRID           6           2.0     -2.0      0.0
GRID           7           2.0      2.0      0.0
GRID           8          -2.0      2.0      0.0
GRID           9          -2.0     -2.0      0.0
GRID          10          -5.0      0.0      0.0
GRID          11          -2.0      0.0      0.0
GRID          12           2.0      0.0      0.0
GRID          13           5.0      0.0      0.0
GRID          14           0.0     -5.0      0.0
GRID          15           0.0     -2.0      0.0
GRID          16           0.0      2.0      0.0
GRID          17           0.0      5.0      0.0
$--1---|---2---|---3---|---4---|---5---|---6---|---7---|---8---|---9---|
CTRIA3         1       1       1      12       7      0.0
CTRIA3         2       1       1       7      16    -45.0
CTRIA3         3       1       1      16       8    -90.0
CTRIA3         4       1       1       8      11   -135.0
CTRIA3         5       1       1      11       9    180.0
CTRIA3         6       1       1       9      15    135.0
CTRIA3         7       1       1      15       6     90.0
CTRIA3         8       1       1       6      12     45.0
```

```
$--1---|---2---|---3---|---4---|---5---|---6---|---7---|---8---|---9---|
CQUAD4        9       1      12      13       3       7     0.0
CQUAD4       10       1       7       3      17      16   -45.0
CQUAD4       11       1      16      17       4       8   -90.0
CQUAD4       12       1       8       4      10      11  -135.0
CQUAD4       13       1      11      10       5       9   180.0
CQUAD4       14       1       9       5      14      15   135.0
CQUAD4       15       1      15      14       2       6    90.0
CQUAD4       16       1       6       2      13      12    45.0
$--1---|---2---|---3---|---4---|---5---|---6---|---7---|---8---|---9---|
MAT8          1   22.0+6   1.30+6    0.30   0.75+6  0.75+6 0.516+6   0.056
+        -0.3-6   18.0-6            170.0+3 170.0+3  6.50+3  28.0+3  10.0+3
+                    0.0
$--1---|---2---|---3---|---4---|---5---|---6---|---7---|---8---|---9---|
PCOMPG        1                      10.0+3    TSAI
              1       1    0.01    -45.0     YES
              2       1    0.01      0.0     YES
              3       1    0.01     45.0     YES
              4       1    0.01     90.0     YES
              5       1    0.01     90.0     YES
              6       1    0.01     45.0     YES
              7       1    0.01      0.0     YES
              8       1    0.01    -45.0     YES
$--1---|---2---|---3---|---4---|---5---|---6---|---7---|---8---|---9---|
SPC           1       4     123      0.0
SPC           1       5      13      0.0
SPC           1       3       3      0.0
$--1---|---2---|---3---|---4---|---5---|---6---|---7---|---8---|---9---|
FORCE         2      10       0    1.0-12000.0     0.0     0.0
FORCE         2       4       0    1.0 -6000.0     0.0     0.0
FORCE         2       5       0    1.0 -6000.0     0.0     0.0
FORCE         2       3       0    1.0  6000.0     0.0     0.0
FORCE         2       2       0    1.0  6000.0     0.0     0.0
FORCE         2      13       0    1.0 12000.0     0.0     0.0
$--1---|---2---|---3---|---4---|---5---|---6-=-|  -7---|---8---|---9---|
TEMPD         3    75.0
TEMPD         4   175.0
ENDDATA
```

OptiStruct Composite Ply-Based Shell Model

```
TITLE = 10" x 10" Plate
SUBTITLE = Ply-Based Shell Model
$Output File Formats
OUTPUT,H3D
OUTPUT,OPTI
$Global Output Requests
DISPLACEMENT = ALL
CSTRAIN(MECH,THER) = ALL
CSTRESS = ALL
STRESS = NO
$Global Subcase Control
TEMPERATURE(INITIAL) = 4
$Subcase Definitions
SUBCASE 1
  LABEL Nx_step
  ANALYSIS STATICS
  SPC = 1
  LOAD = 2
SUBCASE 2
  LABEL Nxt_step
  ANALYSIS STATICS
  SPC = 1
  TEMPERATURE(LOAD) = 3
SUBCASE 3
  LABEL Nx_Nxt_step
  ANALYSIS STATICS
  SPC = 1
  LOAD = 2
  TEMPERATURE(LOAD) = 3
BEGIN BULK
$--1---|---2---|---3---|---4---|---5---|---6---|---7---|---8---|---9---|
GRID           1            0.0     0.0     0.0
GRID           2            5.0    -5.0     0.0
GRID           3            5.0     5.0     0.0
GRID           4           -5.0     5.0     0.0
GRID           5           -5.0    -5.0     0.0
GRID           6            2.0    -2.0     0.0
GRID           7            2.0     2.0     0.0
GRID           8           -2.0     2.0     0.0
GRID           9           -2.0    -2.0     0.0
GRID          10           -5.0     0.0     0.0
GRID          11           -2.0     0.0     0.0
GRID          12            2.0     0.0     0.0
GRID          13            5.0     0.0     0.0
GRID          14            0.0    -5.0     0.0
GRID          15            0.0    -2.0     0.0
GRID          16            0.0     2.0     0.0
GRID          17            0.0     5.0     0.0
$--1---|---2---|---3---|---4---|---5---|---6---|---7---|---8---|---9---|
CTRIA3         1      1      1     12      7      0.0
CTRIA3         2      1      1      7     16    -45.0
CTRIA3         3      1      1     16      8    -90.0
CTRIA3         4      1      1      8     11   -135.0
CTRIA3         5      1      1     11      9    180.0
CTRIA3         6      1      1      9     15    135.0
CTRIA3         7      1      1     15      6     90.0
CTRIA3         8      1      1      6     12     45.0
$--1---|---2---|---3---|---4---|---5---|---6---|---7---|---8---|---9---|
CQUAD4         9      1     12     13      3      7      0.0
CQUAD4        10      1      7      3     17     16    -45.0
CQUAD4        11      1     16     17      4      8    -90.0
CQUAD4        12      1      8      4     10     11   -135.0
CQUAD4        13      1     11     10      5      9    180.0
CQUAD4        14      1      9      5     14     15    135.0
CQUAD4        15      1     15     14      2      6     90.0
CQUAD4        16      1      6      2     13     12     45.0
```

```
$--1---|---2---|---3---|---4---|---5---|---6---|---7---|---8---|---9---|
SET            1    ELEM    LIST
+              1    THRU      16
$--1---|---2---|---3---|---4---|---5---|---6---|---7---|---8---|---9---|
MAT8           1  22.0+6  1.30+6    0.30  0.75+6  0.75+6 0.516+6   0.056
+         -0.3-6  18.0-6          170.0+3 170.0+3  6.50+3  28.0+3  10.0+3
+                    0.0
$--1---|---2---|---3---|---4---|---5---|---6---|---7---|---8---|---9---|
PCOMPP         1                  10.0+3    TSAI
$--1---|---2---|---3---|---4---|---5---|---6---|---7---|---8---|---9---|
PLY            1       1    0.01   -45.0     YES
+              1
PLY            2       1    0.01     0.0     YES
+              1
PLY            3       1    0.01    45.0     YES
+              1
PLY            4       1    0.01    90.0     YES
+              1
PLY            5       1    0.01    90.0     YES
+              1
PLY            6       1    0.01    45.0     YES
+              1
PLY            7       1    0.01     0.0     YES
+              1
PLY            8       1    0.01   -45.0     YES
+              1
$--1---|---2---|---3---|---4---|---5---|---6---|---7---|---8---|---9---|
STACK          1               1       2       3       4       5       6
+              7       8
$--1---|---2---|---3---|---4---|---5---|---6---|---7---|---8---|---9---|
SPC            1       4     123     0.0
SPC            1       5      13     0.0
SPC            1       3       3     0.0
$--1---|---2---|---3---|---4---|   5---|---6---|---7---|---8---|---9---|
FORCE          2      10       0     1.0-12000.0     0.0     0.0
FORCE          2       4       0     1.0 -6000.0     0.0     0.0
FORCE          2       5       0     1.0 -6000.0     0.0     0.0
FORCE          2       3       0     1.0  6000.0     0.0     0.0
FORCE          2       2       0     1.0  6000.0     0.0     0.0
FORCE          2      13       0     1.0 12000.0     0.0     0.0
$--1---|---2---|---3---|---4---|---5---|---6---|---7---|---8---|---9---|
TEMPD          3    75.0
TEMPD          4   175.0
ENDDATA
```

OptiStruct Composite Ply-by-Ply Solid Model

```
TITLE = 10" x 10" Plate
SUBTITLE = Ply-by-Ply Solid Model
$Output File Formats
OUTPUT,H3D
OUTPUT,OPTI
$Global Output Requests
DISPLACEMENT = ALL
STRAIN(MECH,THER) = ALL
STRESS = ALL
$Global Subcase Control
TEMPERATURE(INITIAL) = 4
$Subcase Definitions
SUBCASE 1
  LABEL Nx_step
  ANALYSIS STATICS
  SPC = 1
  LOAD = 2
SUBCASE 2
  LABEL Nxt_step
  ANALYSIS STATICS
  SPC = 1
  TEMPERATURE(LOAD) = 3
SUBCASE 3
  LABEL Nx_Nxt_step
  ANALYSIS STATICS
  SPC = 1
  LOAD = 2
  TEMPERATURE(LOAD) = 3
BEGIN BULK
$--1---|---2---|---3---|---4---|---5---|---6---|---7---|---8---|---9---|
GRID           1              0.0      0.0    -0.04
...
GRID          17              0.0      5.0    -0.04
GRID         101              0.0      0.0    -0.03
...
GRID         117              0.0      5.0    -0.03
GRID         201              0.0      0.0    -0.02
...
GRID         217              0.0      5.0    -0.02
GRID         301              0.0      0.0    -0.01
...
GRID         317              0.0      5.0    -0.01
GRID         401              0.0      0.0     0.0
...
GRID         417              0.0      5.0     0.0
GRID         501              0.0      0.0     0.01
...
GRID         517              0.0      5.0     0.01
GRID         601              0.0      0.0     0.02
...
GRID         617              0.0      5.0     0.02
GRID         701              0.0      0.0     0.03
...
GRID         717              0.0      5.0     0.03
GRID         801              0.0      0.0     0.04
...
GRID         817              0.0      5.0     0.04
```

```
$--1---|---2---|---3---|---4---|---5---|---6---|---7---|---8---|---9---|
CPENTA      101        1      101      107      112        1        7       12
...
CPENTA      108        1      101      112      106        1       12        6
CPENTA      201        2      201      207      212      101      107      112
...
CPENTA      208        2      201      212      206      101      112      106
CPENTA      301        3      301      307      312      201      207      212
...
CPENTA      308        3      301      312      306      201      212      206
CPENTA      401        4      401      407      412      301      307      312
...
CPENTA      408        4      401      412      406      301      312      306
CPENTA      501        5      401      412      407      501      512      507
...
CPENTA      508        5      401      406      412      501      506      512
CPENTA      601        6      501      512      507      601      612      607
...
CPENTA      608        6      501      506      512      601      606      612
CPENTA      701        7      601      612      607      701      712      707
...
CPENTA      708        7      601      606      612      701      706      712
CPENTA      801        8      701      712      707      801      812      807
...
CPENTA      808        8      701      706      712      801      806      812
$--1---|---2---|---3---|---4---|---5---|---6---|---7---|---8---|---9---|
CHEXA       109        1      112      107      103      113       12        7
+             3       13

...
CHEXA       116        1      106      112      113      102        6       12
+            13        2
CHEXA       209        2      212      207      203      213      112      107
+           103      113

...
CHEXA       216        2      206      212      213      202      106      112
+           113      102
CHEXA       309        3      312      307      303      313      212      207
+           203      213

...
CHEXA       316        3      306      312      313      302      206      212
+           213      202
CHEXA       409        4      412      407      403      413      312      307
+           303      313

...
CHEXA       416        4      406      412      413      402      306      312
+           313      302
CHEXA       509        5      412      413      403      407      512      513
+           503      507

...
CHEXA       516        5      406      402      413      412      506      502
+           513      512
CHEXA       609        6      512      513      503      507      612      613
+           603      607

...
CHEXA       616        6      506      502      513      512      606      602
+           613      612
CHEXA       709        7      612      613      603      607      712      713
+           703      707

...
CHEXA       716        7      606      602      613      612      706      702
+           713      712
CHEXA       809        8      712      713      703      707      812      813
+           803      807

...
CHEXA       816        8      706      702      713      712      806      802
+           813      812
```

```
$--1---|---2---|---3---|---4---|---5---|---6---|---7---|---8---|---9---|
CORD2R      1                  0.0     0.0     0.0     0.0     0.0     1.0
+         0.707  -0.707     0.0
CORD2R      2                  0.0     0.0     0.0     0.0     0.0     1.0
+         1.0     0.0      0.0
CORD2R      3                  0.0     0.0     0.0     0.0     0.0     1.0
+         0.707   0.707     0.0
CORD2R      4                  0.0     0.0     0.0     0.0     0.0     1.0
+         0.0     1.0      0.0
$--1---|---2---|---3---|---4---|---5---|---6---|---7---|---8---|---9---|
MAT9ORT     1  22.0+6  1.30e6  1.30e6    0.30    0.26  0.0177   0.056
+        0.75e6 0.561e6 0.75e6  -0.3-6  18.0-6  18.0-6
$--1---|---2---|---3---|---4---|---5---|---6---|---7---|---8---|---9---|
PSOLID      1       1       1
PSOLID      2       1       2
PSOLID      3       1       3
PSOLID      4       1       4
PSOLID      5       1       4
PSOLID      6       1       3
PSOLID      7       1       2
PSOLID      8       1       1
$--1---|---2---|---3---|---4---|---5---|---6---|---7---|---8---|---9---|
SPC         1     404     123     0.0
SPC         1     405      13     0.0
SPC         1     403       3     0.0
$--1---|---2---|---3---|---4---|---5---|---6---|---7---|---8---|---9---|
PLOAD4      2     113 30000.0                             105      10
+           0    -1.0     0.0     0.0
...
PLOAD4      2     116 30000.0                             113       2
+           0     1.0     0.0     0.0
$--1---|---2---|---3---|---4---|---5---|---6---|---7---|---8---|---9---|
TEMPD       3    75.0
TEMPD       4   175.0
ENDDATA
```

Appendix F
OptiStruct Composite Optimization Models

OptiStruct Composite Free-Size Model

```
TITLE = Plate Hole
SUBTITLE = Composite Free-Size Optimization
$Output File Formats
OUTPUT,H3D,FL
OUTPUT,FSTOSZ
$Global Output Requests
DISPLACEMENT = ALL
CSTRAIN(MECH) = ALL
CSTRESS = ALL
THICKNESS=ALL
STRESS = NO
$Global Subcase Control
TEMPERATURE(INITIAL) = 4
DESGLB 1
$Subcase Definitions
SUBCASE 1
  LABEL Nx_Nxt
  ANALYSIS STATICS
  SPC = 1
  LOAD = 2
  TEMPERATURE(LOAD) = 3
  DESOBJ(MIN)= 2
BEGIN BULK
$--1---|---2---|---3---|---4---|---5---|---6---|---7---|---8---|---9---|
GRID           1              -2.0     2.0     0.0
...
GRID        2348            0.093  -0.387     0.0
$--1---|---2---|---3---|---4---|---5---|---6---|---7---|---8---|---9---|
CQUAD4         1       1     111     110     114     113 -54.120
...
CQUAD4      2240       1    2348    2345    2342    2343 168.650
$--1---|---2---|---3---|---4---|---5---|---6---|---7---|---8---|---9---|
SET            1    ELEM    LIST
+              1    THRU    2240
SET            2    ELEM    LIST
+              1    THRU    2240
SET            3    ELEM    LIST
+              1    THRU    2240
SET            4    ELEM    LIST
+              1    THRU    2240
$--1---|---2---|---3---|---4---|---5---|---6---|---7---|---8---|---9---|
MAT8           1  22.0+6  1.30+6    0.30  0.75+6  0.75+6 0.516+6    0.056
+         -0.3-6  18.0-6          170.0+3 170.0+3  6.50+3  28.0+3   10.0+3
+                   0.0
$--1---|---2---|---3---|---4---|---5---|---6---|---7---|---8---|---9---|
PCOMPP         1
$--1---|---2---|---3---|---4---|---5---|---6---|---7---|---8---|---9---|
PLY            1       1     0.1     0.0     YES
+              1
PLY            2       1     0.1    90.0     YES
+              2
PLY            3       1     0.1    45.0     YES
+              3
PLY            4       1     0.1   -45.0     YES
+              4
$--1---|---2---|---3---|---4---|---5---|---6---|---7---|---8---|---9---|
STACK          1   SMEAR       1       2       3       4
```

```
$--1---|---2---|---3---|---4---|---5---|---6---|---7---|---8---|---9---|
DSIZE          1    STACK      1
+            COMP  LAMTHK   0.04
+            COMP  PLYPCT      0     0.2    0.65   BYANG
+            COMP  BALANCE    45     -45            BYANG
+            COMP  PLYDRP    ALL  PLYSLP    0.33
$--1---|---2---|---3---|---4---|---5---|---6---|---7---|---8---|---9---|
DRESP1         1    vfrac VOLFRAC
DRESP1         2    compl   COMP
$--1---|---2---|---3---|---4---|---5---|---6---|---7---|---8---|---9---|
DCONSTR        1      1            0.3
$--1---|---2---|---3---|---4---|---5---|---6---|---7---|---8---|---9---|
SPC            1    405       3    0.0
SPC            1   1041      13    0.0
SPC            1   1061     123    0.0
$--1---|---2---|---3---|---4---|---5---|---6---|---7---|---8---|---9---|
FORCE          2   1042       0    1.0  -560.0     0.0     0.0
...
FORCE          2    413       0    1.0   560.0     0.0     0.0
$--1---|---2---|---3---|---4---|---5---|---6---|---7---|---8---|---9---|
TEMPD          3   75.0
TEMPD          4  175.0
ENDDATA
```

OptiStruct Composite Size Model

```
TITLE = Plate Hole
SUBTITLE = Composite Size Optimization
$Output File Formats
OUTPUT,H3D,FL
OUTPUT,SZTOSH
$Global Output Requests
CSTRAIN(MECH) = ALL
CSTRESS = ALL
DISPLACEMENT = ALL
THICKNESS= ALL
STRESS = NO
$Global Subcase Control
TEMPERATURE(INITIAL) = 4
DESOBJ(MIN)= 1
$Subcase Definitions
SUBCASE 1
  LABEL Nx_Nxt
  ANALYSIS STATICS
  SPC = 1
  LOAD = 2
  TEMPERATURE(LOAD) = 3
  DESSUB = 3
BEGIN BULK
$--1---|---2---|---3---|---4---|---5---|---6---|---7---|---8---|---9---|
GRID           1              -2.0     2.0     0.0
...
GRID        2348              0.093  -0.387    0.0
$--1---|---2---|---3---|---4---|---5---|---6---|---7---|---8---|---9---|
CQUAD4         1         1     111     110     114     113 -54.120
...
CQUAD4      2240         1    2348    2345    2342    2343 168.650
$--1---|---2---|---3---|---4---|---5---|---6---|---7---|---8---|---9---|
SET        11100     ELEM    LIST
+              1     THRU    2240
SET        11200     ELEM    LIST
+              1        2       3       4       5       6       7       8
...
SET        11300     ELEM    LIST
+              1        3       5       6       7       8       9      11
...
SET        11400     ELEM    LIST
+            161      162     163     164     165     166     167     168
...
SET        12100     ELEM    LIST
+              1     THRU    2240
SET        12200     ELEM    LIST
+              1        3       5       6       7       8       9      11
...
SET        12300     ELEM    LIST
+            161      162     163     164     165     166     167     168
...
SET        12400     ELEM    LIST
+            241      242     243     244     245     246     247     248
...
SET        13100     ELEM    LIST
+              1     THRU    2240
SET        13200     ELEM    LIST
+             41       43      45      46      47      48      53      59
...
SET        13300     ELEM    LIST
+            241      242     243     244     245     246     247     248
...
SET        13400     ELEM    LIST
+           1851     1852    1853    1854    1855    1856    1857    1858
...
```

```
$--1---|---2---|---3---|---4---|---5---|---6---|---7---|---8---|---9---|
MAT8           1   22.0+6  1.30+6     0.30  0.75+6  0.75+6 0.516+6    0.056
+         -0.3-6  18.0-6            170.0+3 170.0+3  6.50+3  28.0+3   10.0+3
+                    0.0
$--1---|---2---|---3---|---4---|---5---|---6---|---7---|---8---|---9---|
FCOMPP         1                    10.0+3    TSAI
$--1---|---2---|---3---|---4---|---5---|---6---|---7---|---8---|---9---|
PLY        11100       1    0.04     0.0       YES    0.01
+          11100
PLY        11200       1    0.04     0.0       YES    0.01
+          11200
PLY        11300       1    0.04     0.0       YES    0.01
+          11300
PLY        11400       1    0.04     0.0       YES    0.01
+          11400
PLY        12100       1    0.04    90.0       YES    0.01
+          12100
PLY        12200       1    0.04    90.0       YES    0.01
+          12200
PLY        12300       1    0.04    90.0       YES    0.01
+          12300
PLY        12400       1    0.04    90.0       YES    0.01
+          12400
PLY        13100       1    0.04    45.0       YES    0.01
+          13100
PLY        13200       1    0.04    45.0       YES    0.01
+          13200
PLY        13300       1    0.04    45.0       YES    0.01
+          13300
PLY        13400       1    0.04    45.0       YES    0.01
+          13400
PLY        14100       1    0.04   -45.0       YES    0.01
+          13100
PLY        14200       1    0.04    45.0       YES    0.01
+          13200
PLY        14300       1    0.04   -45.0       YES    0.01
+          13300
PLY        14400       1    0.04   -45.0       YES    0.01
+          13400
$--1---|---2---|---3---|---4---|---5---|---6---|---7---|---8---|---9---|
STACK          1 SYSMEAR   11100   12100   13100   14100   11200   12200
+          13200   14200   11300   12300   13300   14300   11400   12400
+          13400   14400
$--1---|---2---|---3---|---4---|---5---|---6---|---7---|---8---|---9---|
DESVAR     11100  D11100    0.04     0.0       0.2
DESVAR     11200  D11200    0.04     0.0       0.2
DESVAR     11300  D11300    0.04     0.0       0.2
DESVAR     11400  D11400    0.04     0.0       0.2
DESVAR     12100  D12100    0.04     0.0       0.2
DESVAR     12200  D12200    0.04     0.0       0.2
DESVAR     12300  D12300    0.04     0.0       0.2
DESVAR     12400  D12400    0.04     0.0       0.2
DESVAR     13100  D13100    0.04     0.0       0.2
DESVAR     13200  D13200    0.04     0.0       0.2
DESVAR     13300  D13300    0.04     0.0       0.2
DESVAR     13400  D13400    0.04     0.0       0.2
DESVAR     14100  D14100    0.04     0.0       0.2
DESVAR     14200  D14200    0.04     0.0       0.2
DESVAR     14300  D14300    0.04     0.0       0.2
DESVAR     14400  D14400    0.04     0.0       0.2
$--1---|---2---|---3---|---4---|---5---|---6---|---7---|---8---|---9---|
DCOMP          1   STACK       1
+         LAMTHK    0.04
+         PLYPCT       0     .20     .65   BYANG
+         BALANCE     45     -45           BYANG
```

```
$--1---|---2---|---3---|---4---|---5---|---6---|---7---|---8---|---9---|
DVPREL1   11100      PLY   11100      T                        0.0
+         11100      1.0
DVPREL1   11200      PLY   11200      T                        0.0
+         11200      1.0
DVPREL1   11300      PLY   11300      T                        0.0
+         11300      1.0
DVPREL1   11400      PLY   11400      T                        0.0
+         11400      1.0
DVPREL1   12100      PLY   12100      T                        0.0
+         12100      1.0
DVPREL1   12200      PLY   12200      T                        0.0
+         12200      1.0
DVPREL1   12300      PLY   12300      T                        0.0
+         12300      1.0
DVPREL1   12400      PLY   12400      T                        0.0
+         12400      1.0
DVPREL1   13100      PLY   13100      T                        0.0
+         13100      1.0
DVPREL1   13200      PLY   13200      T                        0.0
+         13200      1.0
DVPREL1   13300      PLY   13300      T                        0.0
+         13300      1.0
DVPREL1   13400      PLY   13400      T                        0.0
+         13400      1.0
DVPREL1   14100      PLY   14100      T                        0.0
+         14100      1.0
DVPREL1   14200      PLY   14200      T                        0.0
+         14200      1.0
DVPREL1   14300      PLY   14300      T                        0.0
+         14300      1.0
DVPREL1   14400      PLY   14400      T                        0.0
+         14400      1.0
$--1---|---2---|---3---|---4---|---5---|---6---|---7---|---8---|---9---|
DRESP1        1    mass    MASS
DRESP1        2    fiber CSTRAIN     PLY              MS1
DRESP1        3    plyCFAILURE       PLY              TSAI
$--1---|---2---|---3---|---4---|---5---|---6---|---7---|---8---|---9---|
DCONSTR       1      2           0.003
DCONSTR       2      3           1.0
$--1---|---2---|---3---|---4---|---5---|---6---|---7---|---8---|---9---|
DCONADD       3      1      2
$--1---|---2---|---3---|---4---|---5---|---6---|---7---|---8---|---9---|
SPC           1    405      3    0.0
SPC           1   1041     13    0.0
SPC           1   1061    123    0.0
$--1---|---2---|---3---|---4---|---5---|---6---|---7---|---8---|---9---|
FORCE         2   1042      0    1.0  -560.0     0.0      0.0
...
FORCE         2    413      0    1.0   560.0     0.0      0.0
$--1---|---2---|---3---|---4---|---5---|---6---|---7---|---8---|---9---|
TEMPD         3   75.0
TEMPD         4  175.0
ENDDATA
```

OptiStruct Composite Shuffling Model

```
TITLE = Plate Hole
SUBTITLE = Composite Shuffling Optimization
$Output File Formats
OUTPUT,H3D,FL
$Global Output Requests
CSTRAIN(MECH) = ALL
CSTRESS = ALL
DISPLACEMENT = ALL
STRESS = NO
$Global Subcase Control
TEMPERATURE(INITIAL) = 4
DESOBJ(MIN)= 1
$Subcase Definitions
SUBCASE 1
  LABEL Nx_Nxt
  ANALYSIS STATICS
  SPC = 1
  LOAD = 2
  TEMPERATURE(LOAD) = 3
  DESSUB = 3
BEGIN BULK
$--1---|---2---|---3---|---4---|---5---|---6---|---7---|---8---|---9---|
GRID           1              -2.0     2.0      0.0
...
GRID        2348              0.093  -0.387     0.0
$--1---|---2---|---3---|---4---|---5---|---6---|---7---|---8---|---9---|
CQUAD4         1       1       111     110      114      113 -54.120
...
CQUAD4      2240       1      2348    2345     2342     2343 168.650
$--1---|---2---|---3---|---4---|---5---|---6---|---7---|---8---|---9---|
SET        11100    ELEM    LIST
+              1    THRU    2240
SET        11200    ELEM    LIST
+              1       2       3       4        5        6        7        8
...
SET        12100    ELEM    LIST
+              1    THRU    2240
SET        12400    ELEM    LIST
+            241     242     243     244      245      246      247      248
...
SET        13100    ELEM    LIST
+              1    THRU    2240
SET        13300    ELEM    LIST
+            241     242     243     244      245      246      247      248
...
SET        13400    ELEM    LIST
+           1851    1852    1853    1854     1855     1856     1857     1858
...
$--1---|---2---|---3---|---4---|---5---|---6---|---7---|---8---|---9---|
MAT8           1   22.0+6  1.30+6    0.30   0.75+6   0.75+6 0.516+6    0.056
+          -0.3-6  18.0-6          170.0+3  170.0+3  6.50+3  28.0+3   10.0+3
+             0.0
$--1---|---2---|---3---|---4---|---5---|---6---|---7---|---8---|---9---|
PCOMPP         1                   10.0+3    TSAI
```

```
$--1---|---2---|---3---|---4---|---5---|---6---|-  7---|---8---|---9---|
PLY        11101      1   0.01    0.0      YES
+          11100
PLY        11102      1   0.01    0.0      YES
+          11100
PLY        11201      1   0.01    0.0      YES
+          11200
PLY        11202      1   0.01    0.0      YES
+          11200
PLY        12101      1   0.01   90.0      YES
+          12100
PLY        12401      1   0.01   90.0      YES
+          12400
PLY        12402      1   0.01   90.0      YES
+          12400
PLY        12403      1   0.01   90.0      YES
+          12400
PLY        13101      1   0.01   45.0      YES
+          13100
PLY        13301      1   0.01   45.0      YES
+          13300
PLY        13302      1   0.01   45.0      YES
+          13300
PLY        13401      1   0.01   45.0      YES
+          13400
PLY        13402      1   0.01   45.0      YES
+          13400
PLY        14101      1   0.01  -45.0      YES
+          13100
PLY        14301      1   0.01  -45.0      YES
+          13300
PLY        14302      1   0.01  -45.0      YES
+          13300
PLY        14401      1   0.01  -45.0      YES
+          13400
PLY        14402      1   0.01  -45.0      YES
+          13400
$--1---|---2---|---3---|---4---|---5---|---6---|---7---|---8---|---9---|
STACK          1     SYM  11101   11102   12101   13101   14101   11201
+                  11202   13301   13302   14301   14302   12401   12402
+                  12403   13401   13402   14401   14402
$--1---|---2---|---3---|---4---|---5---|---6---|---7---|---8---|---9---|
DSHUFFLE       1   STACK       1
           MAXSUCC   ALL       1       0
             COVER     1     -45       0      45      90
$--1---|---2---|---3---|---4---|---5---|---6---|---7---|---8---|---9---|
DRESP1         1    mass    MASS
DRESP1         2   fiber CSTRAIN     PLY             MS1
DRESP1         3   plyCFAILURE     PLY            TSAI
$--1---|---2---|---3---|---4---|---5---|---6---|---7---|---8---|---9---|
DCONSTR        1       2           0.003
DCONSTR        2       3             1.0
$--1---|---2---|---3---|---4---|---5---|---6---|---7---|---8---|---9---|
DCONADD        3       1       2
$--1---|---2---|---3---|---4---|---5---|---6---|---7---|---8---|---9---|
SPC            1     405       3     0.0
SPC            1    1041      13     0.0
SPC            1    1061     123     0.0
$--1---|---2---|---3---|---4---|---5---|---6---|---7---|---8---|---9---|
FORCE          2    1042       0     1.0  -560.0     0.0     0.0
...
FORCE          2     413       0     1.0   560.0     0.0     0.0
$--1---|---2---|---3---|---4---|---5---|---6---|---7---|---8---|---9---|
TEMPD          3    75.0
TEMPD          4   175.0
ENDDATA
```

OptiStruct Composite Final Design Model

```
TITLE = Plate Hole
SUBTITLE = Composite Final Design
$Output File Formats
OUTPUT,H3D
$Global Output Requests
DISPLACEMENT = ALL
CSTRAIN(MECH) = ALL
CSTRESS = ALL
STRESS = NO
$Global Subcase Control
TEMPERATURE(INITIAL) = 4
$Subcase Definitions
SUBCASE 1
  LABEL Nx_Nxt
  ANALYSIS STATICS
  SPC = 1
  LOAD = 2
  TEMPERATURE(LOAD) = 3
BEGIN BULK
$--1---|---2---|---3---|---4---|---5---|---6---|---7---|---8---|---9---|
GRID           1            -2.0     2.0     0.0
...
GRID        2348           0.093  -0.387     0.0
$--1---|---2---|---3---|---4---|---5---|---6---|---7---|---8---|---9---|
CQUAD4         1       1     111     110     114     113 -54.120
...
CQUAD4      2240       1    2348    2345    2342    2343 168.650
$--1---|---2---|---3---|---4---|---5---|---6---|---7---|---8---|---9---|
SET        11100    ELEM    LIST
+              1    THRU    2240
SET        11200    ELEM    LIST
+              1       2       3       4       5       6       7       8
...
SET        12100    ELEM    LIST
+              1    THRU    2240
SET        12400    ELEM    LIST
+            241     242     243     244     245     246     247     248
...
SET        13100    ELEM    LIST
+              1    THRU    2240
SET        13300    ELEM    LIST
+            241     242     243     244     245     246     247     248
...
SET        13400    ELEM    LIST
+           1851    1852    1853    1854    1855    1856    1857    1858
...
$--1---|---2---|---3---|---4---|---5---|---6---|---7---|---8---|---9---|
MAT8           1  22.0+6  1.30+6    0.30  0.75+6  0.75+6 0.516+6   0.056
+         -0.3-6  18.0-6          170.0+3 170.0+3  6.50+3  28.0+3  10.0+3
+            0.0
$--1---|---2---|---3---|---4---|---5---|---6---|---7---|---8---|---9---|
PCOMPP         1                   10.0+3    TSAI
```

```
$--1---|---2---|---3---|---4---|---5---|---6---|--•7 •-|---8---|---9---|
PLY        11101       1    0.01    0.0      YES
+          11100
PLY        11102       1    0.01    0.0      YES
+          11100
PLY        11201       1    0.01    0.0      YES
+          11200
PLY        11202       1    0.01    0.0      YES
+          11200
PLY        12101       1    0.01    90.0     YES
+          12100
PLY        12401       1    0.01    90.0     YES
+          12400
PLY        12402       1    0.01    90.0     YES
+          12400
PLY        12403       1    0.01    90.0     YES
+          12400
PLY        13101       1    0.01    45.0     YES
+          13100
PLY        13301       1    0.01    45.0     YES
+          13300
PLY        13302       1    0.01    45.0     YES
+          13300
PLY        13401       1    0.01    45.0     YES
+          13400
PLY        13402       1    0.01    45.0     YES
+          13400
PLY        14101       1    0.01   -45.0     YES
+          13100
PLY        14301       1    0.01   -45.0     YES
+          13300
PLY        14302       1    0.01   -45.0     YES
+          13300
PLY        14401       1    0.01   -45.0     YES
+          13400
PLY        14402       1    0.01   -45.0     YES
+          13400
$--1---|---2---|---3---|---4---|---5---|---6---|---7---|---8---|---9---|
STACK       1     SYM   14101   11101   13101   12101   14301   11201
+          13301  12401  14302   11202   13302   12402   14401   11102
+          13401  12403  14402   13402
$--1---|---2---|---3---|---4---|---5---|---6---|---7---|---8---|---9---|
SPC         1     405      3     0.0
SPC         1    1041     13     0.0
SPC         1    1061    123     0.0
$--1---|---2---|---3---|---4---|---5---|---6---|---7---|---8---|---9---|
FORCE       2    1042      0     1.0   -560.0    0.0     0.0
...
FORCE       2     413      0     1.0    560.0    0.0     0.0
$--1---|---2---|---3---|---4---|---5---|---6---|---7---|---8---|---9---|
TEMPD       3    75.0
TEMPD       4   175.0
ENDDATA
```

.

Appendix G
Answers to Chapter Exercises

Chapter 2 Exercise Answers

2.1.

$$v' = \begin{bmatrix} \dfrac{\sqrt{2}}{2} & \dfrac{\sqrt{2}}{2} & 0 \\[2mm] -\dfrac{\sqrt{2}}{2} & \dfrac{\sqrt{2}}{2} & 0 \\[2mm] 0 & 0 & 1 \end{bmatrix} \begin{Bmatrix} 1 \\ 1 \\ 0 \end{Bmatrix} \qquad v' = \begin{Bmatrix} \sqrt{2} \\ 0 \\ 0 \end{Bmatrix}$$

2.2.

$$\{\sigma\}_{x'} = [Ts]\{\sigma\}_x$$

$$\{\sigma\}_{x'} = \begin{bmatrix} 0.5 & 0.5 & 0 & 1 & 0 & 0 \\ 0.5 & 0.5 & 0 & -1 & 0 & 0 \\ 0 & 0 & 1 & 0 & 0 & 0 \\ -0.5 & 0.5 & 0 & 0 & 0 & 0 \\ 0 & 0 & 0 & 0 & 0.707 & -0.707 \\ 0 & 0 & 0 & 0 & 0.707 & 0.707 \end{bmatrix} \{\sigma\} \qquad \{\sigma\}_{x'} = \begin{Bmatrix} 92.85 \\ 17.65 \\ 36.50 \\ 3.75 \\ 10.61 \\ 10.61 \end{Bmatrix} ksi$$

$$\{\varepsilon\}_{x'} = [Te]\{\varepsilon\}_x$$

$$\{\varepsilon\}_{x'} = \begin{bmatrix} 0.5 & 0.5 & 0 & 0.5 & 0 & 0 \\ 0.5 & 0.5 & 0 & -0.5 & 0 & 0 \\ 0 & 0 & 1 & 0 & 0 & 0 \\ -1 & 1 & 0 & 0 & 0 & 0 \\ 0 & 0 & 0 & 0 & 0.707 & -0.707 \\ 0 & 0 & 0 & 0 & 0.707 & 0.707 \end{bmatrix} \{\varepsilon\}_x \qquad \{\varepsilon\}_{x'} = \begin{Bmatrix} 3000 \\ 2000 \\ 0 \\ 1000 \\ 2828 \\ 2828 \end{Bmatrix} \mu\varepsilon$$

2.3.
The strain invariants J_1, J_2, and J_3 and the von Mises strain will be the same for both strain tensors since these values are invariant to strain tensor transformations.

$$J_1 = 5000\mu\varepsilon$$
$$J_2 = 1.75e6\mu\varepsilon$$
$$J_3 = -8.00e9\mu\varepsilon$$
$$\varepsilon_{vm} = 4444\mu\varepsilon$$

2.4.
$$[\overline{C}] = [Ts][C][Te]^{-1}$$
[Ts] from the solution to problem 2.2

$$[Te]^{-1} = \begin{bmatrix} 0.5 & 0.5 & 0 & -0.5 & 0 & 0 \\ 0.5 & 0.5 & 0 & 0.5 & 0 & 0 \\ 0 & 0 & 1 & 0 & 0 & 0 \\ 1 & -1 & 0 & 0 & 0 & 0 \\ 0 & 0 & 0 & 0 & 0.707 & 0.707 \\ 0 & 0 & 0 & 0 & -0.707 & 0.707 \end{bmatrix}$$

$$[\overline{C}] = \begin{bmatrix} 1.48e7 & 7.30e6 & 7.30e6 & 0 & 0 & 0 \\ 7.30e6 & 1.48e7 & 7.30e6 & 0 & 0 & 0 \\ 7.30e6 & 7.30e6 & 1.48e7 & 0 & 0 & 0 \\ 0 & 0 & 0 & 3.76e6 & 0 & 0 \\ 0 & 0 & 0 & 0 & 3.76e6 & 0 \\ 0 & 0 & 0 & 0 & 0 & 3.76e6 \end{bmatrix}$$

Chapter 3 Exercise Answers

3.1.

$$[S] = \begin{bmatrix} 4.55e-8 & -1.36e-8 & -1.39e-8 & 0 & 0 & 0 \\ -1.36e-8 & 7.69e-7 & -2.00e-7 & 0 & 0 & 0 \\ -1.36e-8 & -2.00e-7 & 7.69e-7 & 0 & 0 & 0 \\ 0 & 0 & 0 & 1.33e-6 & 0 & 0 \\ 0 & 0 & 0 & 0 & 1.94e-6 & 0 \\ 0 & 0 & 0 & 0 & 0 & 1.33e-6 \end{bmatrix}$$

$$[C] = \begin{bmatrix} 2.23e7 & 5.35e5 & 5.35e5 & 0 & 0 & 0 \\ 5.35e5 & 1.41e6 & 3.75e5 & 0 & 0 & 0 \\ 5.35e5 & 3.75e5 & 1.41e6 & 0 & 0 & 0 \\ 0 & 0 & 0 & 7.50e5 & 0 & 0 \\ 0 & 0 & 0 & 0 & 5.16e5 & 0 \\ 0 & 0 & 0 & 0 & 0 & 7.50e5 \end{bmatrix}$$

$$[S^*] = \begin{bmatrix} 4.55e-8 & -1.36e-8 & 0 \\ -1.36e-8 & 7.69e-7 & 0 \\ 0 & 0 & 1.33e-6 \end{bmatrix}$$

$$[Q] = \begin{bmatrix} 2.21e7 & 3.92e5 & 0 \\ 3.92e5 & 1.31e6 & 0 \\ 0 & 0 & 7.50e5 \end{bmatrix}$$

3.2.
Given that E_1 = 50,000 psi, start by determining which transversely isotropic engineering constants are approximately zero. The transversely isotropic engineering constants that need to be determined are E_2, v_{12}, v_{23}, and G_{12}. Looking at the figure of the honeycomb core it should be obvious that $E_2 \approx 0$. However since we cannot use exactly zero, we'll use a numerical zero and set $E_2 = 500$ psi. It should also be obvious that $v_{12} \approx 0.3$ and that $v_{23} \approx 1.0$. Using the restrictions on transversely isotropic engineering constants given by equation (3.39), v_{12} and v_{23} can be selected as follows;

$$\left| \upsilon_{12} \right| < \sqrt{\frac{50{,}000}{500}} = 10$$

$$\upsilon_{23} < 1$$

Selecting $v_{12} = 0.2$ and $v_{23} = 0.9$, which seem reasonable to the envisioned behavior, we need to check against the last restriction of equation (2.39).

$$1 - (2)(0.2)^2 \left(\frac{500}{50{,}000} \right) - 0.9^2 - (2)(0.2)^2 (0.9) \left(\frac{500}{50{,}000} \right) = 0.188 > 0$$

Therefore the selected Poisson ratio values are valid. The only remaining transversely isotropic engineering constant to determine is G_{12} which is typically one order of magnitude less than the E_1 stiffness values, therefore we'll select $G_{12} = 5{,}000$ psi. The final transversely isotropic engineering constants are;

$E_1 = 50{,}000$ psi
$E_2 = 500$ psi
$v_{12} = 0.2$
$v_{23} = 0.9$
$G_{12} = 5{,}000$ psi
$G_{23} = 131.6$ psi

3.3.
a) Since the bar is unconstrained on the table, the boundary condition on the bar is given by $\{\sigma\} = 0$. Therefore, the mechanical strain $\{\varepsilon\}_m = 0$. Furthermore, using the definition of mechanical strain, the total strain must equal the free thermal strain. The free thermal strain is equal to, $\{\varepsilon\}_t = \{\alpha\}\Delta T$, and therefore the total strain is equal to $\{\varepsilon\} = \{\alpha\}\Delta T$ also. Summarizing; the stress, total strain, free thermal strain, and mechanical strain are;

$$\{\sigma\} = \begin{Bmatrix} 0 \\ 0 \\ 0 \\ 0 \\ 0 \\ 0 \end{Bmatrix}, \{\varepsilon\} = \begin{Bmatrix} 0.0012 \\ 0.0012 \\ 0.0012 \\ 0 \\ 0 \\ 0 \end{Bmatrix}, \{\varepsilon\}_t = \begin{Bmatrix} 0.0012 \\ 0.0012 \\ 0.0012 \\ 0 \\ 0 \\ 0 \end{Bmatrix}, \{\varepsilon\}_m = \begin{Bmatrix} 0 \\ 0 \\ 0 \\ 0 \\ 0 \\ 0 \end{Bmatrix}$$

b) Since the bar is completely constrained on the table, the boundary condition on the bar is given by $\{\varepsilon\} = 0$. Therefore, the mechanical strain must be equal to minus the free thermal strain. The free thermal strain is equal to, $\{\varepsilon\}_t = \{\alpha\}\Delta T$, and therefore the mechanical strain is equal to, $\{\varepsilon\}_m = -\{\alpha\}\Delta T$. Once the strains have been determined, the stress $\{\sigma\}$ in the bar can be determined. Summarizing; the total strain, free thermal strain, mechanical strain, and stress are;

$$\{\varepsilon\} = \begin{Bmatrix} 0 \\ 0 \\ 0 \\ 0 \\ 0 \\ 0 \end{Bmatrix}, \{\varepsilon\}_t = \begin{Bmatrix} 0.0012 \\ 0.0012 \\ 0.0012 \\ 0 \\ 0 \\ 0 \end{Bmatrix}, \{\varepsilon\}_m = \begin{Bmatrix} -0.0012 \\ -0.0012 \\ -0.0012 \\ 0 \\ 0 \\ 0 \end{Bmatrix}, \{\sigma\} = \begin{Bmatrix} -35,300 \\ -35,300 \\ -35,300 \\ 0 \\ 0 \\ 0 \end{Bmatrix}$$

The aluminum isotropic stiffness matrix is;

$$[C] = \begin{bmatrix} 1.48e7 & 7.30e6 & 7.30e6 & 0 & 0 & 0 \\ 7.30e6 & 1.48e7 & 7.30e6 & 0 & 0 & 0 \\ 7.30e6 & 7.30e6 & 1.48e7 & 0 & 0 & 0 \\ 0 & 0 & 0 & 3.76e6 & 0 & 0 \\ 0 & 0 & 0 & 0 & 3.76e6 & 0 \\ 0 & 0 & 0 & 0 & 0 & 3.76e6 \end{bmatrix}$$

3.4.

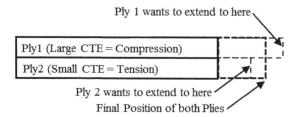

Ply 1 wants to extend to here

Ply1 (Large CTE = Compression)

Ply2 (Small CTE = Tension)

Ply 2 wants to extend to here

Final Position of both Plies

Chapter 4 Exercise Answers

4.1.

First the transversely isotropic shear modulus in the 23-plane for the fiber must be calculated.

$G_{23}^f = 1.00e6$ psi

Second, the isotropic shear modulus for the matrix must be calculated.

$G^m = 0.212e6$ psi

Finally, the table below summarizes the calculated homogenized engineering constants for the rule of mixtures and modified rule of mixtures theories.

	Rule of Mixtures	Modified Rule of Mixtures
E_1	22.12e6	22.12e6
E_2	1.02e6	1.30e6
ν_{12}	0.30	0.30
ν_{23}	0.27	0.26
G_{12}	0.50e6	0.75e6
G_{23}	0 40e6	0.52e6
α_1	-0.29e-6	-0.29e 6
α_2	18.00e-6	18.00e-6

4.2.

Since E_1 is determined by a simple rule of mixtures approach, which is accurate to measured data, simply add the third material volume fraction into the equation.

$$E_1 = E_{f1}V_{f1} + E_{f2}V_{f2} + E_m V_m$$
$$V_{f1} + V_{f2} + V_m = 1$$

4.3.
Since E_2 behaves as a set of springs in series and the fiber is stiffer than the matrix, the total deformation of the fiber-matrix system (top springs) will be less than if all the springs were matrix (bottom springs). Therefore, the transverse modulus E_2 will be stiffer than the matrix modulus E_m.

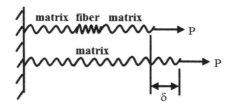

4.4.
To solve this problem, first calculate the transversely isotropic stiffness matrix of the fiber material $[C]^f$

$$[C]^f = \begin{bmatrix} 3.70e7 & 9.14e5 & 9.14e5 & 0 & 0 & 0 \\ 9.14e5 & 2.52e6 & 5.23e5 & 0 & 0 & 0 \\ 9.14e5 & 5.23e5 & 2.52e6 & 0 & 0 & 0 \\ 0 & 0 & 0 & 6.0e6 & 0 & 0 \\ 0 & 0 & 0 & 0 & 1.00e6 & 0 \\ 0 & 0 & 0 & 0 & 0 & 6.0e6 \end{bmatrix}$$

Then calculate the isotropic stiffness matrix of the matrix material $[C]^m$

$$[C]^m = \begin{bmatrix} 7.40e5 & 3.17e5 & 3.17e5 & 0 & 0 & 0 \\ 3.17e5 & 7.40e5 & 3.17e5 & 0 & 0 & 0 \\ 3.17e5 & 3.17e5 & 7.40e5 & 0 & 0 & 0 \\ 0 & 0 & 0 & 2.12e5 & 0 & 0 \\ 0 & 0 & 0 & 0 & 2.12e5 & 0 \\ 0 & 0 & 0 & 0 & 0 & 2.12e5 \end{bmatrix}$$

Then calculate the transversely isotropic stiffness matrix of the homogenized ply material $[C]$. The homogenized engineering constants for the ply were calculated in problem 4.1 as the Modified Rule of Mixtures values.

$$[C] = \begin{bmatrix} 2.23e7 & 5.35e5 & 5.35e5 & 0 & 0 & 0 \\ 5.35e5 & 1.41e6 & 3.75e5 & 0 & 0 & 0 \\ 5.35e5 & 3.75e5 & 1.41e6 & 0 & 0 & 0 \\ 0 & 0 & 0 & 7.50e5 & 0 & 0 \\ 0 & 0 & 0 & 0 & 5.16e5 & 0 \\ 0 & 0 & 0 & 0 & 0 & 7.50e5 \end{bmatrix}$$

Finally calculate the coefficient of thermal expansion vectors for the fiber, matrix, and ply materials respectively. Now using these values you can solve the problem as below.

$$\{\alpha\}^f = \begin{Bmatrix} -0.60e-6 \\ 4.0e-6 \\ 4.0e-6 \\ 0 \\ 0 \\ 0 \end{Bmatrix}, \quad \{\alpha\}^m = \begin{Bmatrix} 30.0e-6 \\ 30.0e-6 \\ 30.0e-6 \\ 0 \\ 0 \\ 0 \end{Bmatrix}, \quad \{\alpha\} = \begin{Bmatrix} -0.30e-6 \\ 18.0e-6 \\ 18.0e-6 \\ 0 \\ 0 \\ 0 \end{Bmatrix}$$

a) The fiber mechanical strain amplification matrix $[M]^f$ and thermal superposition vector $\{T\}^f$ are; $(V^f = 0.60)$

$$[M]^f = \begin{bmatrix} 0.99 & 0 & 0 & 0 & 0 & 0 \\ -0.12 & 0.63 & -0.02 & 0 & 0 & 0 \\ -0.12 & -0.02 & 0.63 & 0 & 0 & 0 \\ 0 & 0 & 0 & 0.16 & 0 & 0 \\ 0 & 0 & 0 & 0 & 0.64 & 0 \\ 0 & 0 & 0 & 0 & 0 & 0.16 \end{bmatrix}$$

$$\{T\}^f = \begin{Bmatrix} -3.04e-7 \\ 1.07e-7 \\ 1.07e-7 \\ 0 \\ 0 \\ 0 \end{Bmatrix}$$

b) The matrix mechanical strain amplification matrix $[M]^m$ and thermal superposition vector $\{T\}^m$ are; ($V^f = 0.60$)

$$[M]^m = \begin{bmatrix} 1.01 & 0 & 0 & 0 & 0 & 0 \\ 0.17 & 1.56 & 0.03 & 0 & 0 & 0 \\ 0.17 & 0.03 & 1.56 & 0 & 0 & 0 \\ 0 & 0 & 0 & 2.27 & 0 & 0 \\ 0 & 0 & 0 & 0 & 1.54 & 0 \\ 0 & 0 & 0 & 0 & 0 & 2.27 \end{bmatrix}$$

$$\{T\}^m = \begin{Bmatrix} 3.03e-5 \\ -9.16e-6 \\ -9.16e-6 \\ 0 \\ 0 \\ 0 \end{Bmatrix}$$

4.5.

a) The average mechanical strain tensor in the fiber with $\{\varepsilon\}_m = \{1\}$ and $\Delta T = 0$ is;.

$$\{\varepsilon\}_m^f = \begin{Bmatrix} 0.99 \\ 0.49 \\ 0.49 \\ 0.16 \\ 0.64 \\ 0.16 \end{Bmatrix}$$

b) The average mechanical strain tensor in the matrix with $\{\varepsilon\}_m = \{1\}$ and $\Delta T = 0$ is;

$$\{\varepsilon\}_m^m = \begin{Bmatrix} 1.01 \\ 1.76 \\ 1.76 \\ 2.27 \\ 1.54 \\ 2.27 \end{Bmatrix}$$

4.6.

a) The average mechanical strain tensor in the fiber with $\{\varepsilon\}_m = \{0\}$ and $\Delta T = -100°F$ is;

$$\{\varepsilon\}_m^f = \begin{Bmatrix} -30.4 \\ 10.7 \\ 10.7 \\ 0 \\ 0 \\ 0 \end{Bmatrix} \mu\varepsilon$$

b) The average mechanical strain tensor in the matrix with $\{\varepsilon\}_m = \{0\}$ and $\Delta T = -100°F$ is;

$$\{\varepsilon\}_m^m = \begin{Bmatrix} 3030.7 \\ -916.1 \\ -916.1 \\ 0 \\ 0 \\ 0 \end{Bmatrix} \mu\varepsilon$$

4.7.

The mechanical strain amplification matrix for the fiber $[M]^f$ accounts for the strain in the fiber due to the stiffness difference between the fiber and matrix constituents. The mechanical strain amplification matrix for the matrix $[M]^m$ accounts for the strain in the matrix due to the stiffness difference between the fiber and matrix constituents.

4.8.

The thermal superposition vector for the fiber $\{T\}^f$ accounts for the strain in the fiber due to the coefficient of thermal expansion difference between the fiber and matrix constituents. The thermal superposition vector for the matrix $\{T\}^m$ accounts for the strain in the matrix due to the coefficient of thermal expansion difference between the fiber and matrix constituents.

Chapter 5 Exercise Answers

5.1.
See figure 5.2. Selected answers for various θ values are given below;

$$[\overline{Q}]_0 = \begin{bmatrix} 2.21e7 & 3.92e5 & 0 \\ 3.92e5 & 1.31e6 & 0 \\ 0 & 0 & 7.50e5 \end{bmatrix}$$

$$[\overline{Q}]_{30} = \begin{bmatrix} 1.32e7 & 4.07e6 & 6.63e6 \\ 4.07e6 & 2.83e6 & 2.38e6 \\ 6.63e6 & 2.38e6 & 4.43e6 \end{bmatrix}$$

$$[\overline{Q}]_{45} = \begin{bmatrix} 6.80e6 & 5.30e6 & 5.20e6 \\ 5.30e6 & 6.80e6 & 5.20e6 \\ 5.20e6 & 5.20e6 & 5.66e6 \end{bmatrix}$$

$$[\overline{Q}]_{60} = \begin{bmatrix} 2.83e6 & 4.07e6 & 2.38e6 \\ 4.07e6 & 1.32e7 & 6.63e6 \\ 2.38e6 & 6.63e6 & 4.43e6 \end{bmatrix}$$

$$[\overline{Q}]_{90} = \begin{bmatrix} 1.31e6 & 3.92e5 & 0 \\ 3.92e5 & 2.21e7 & 0 \\ 0 & 0 & 7.50e5 \end{bmatrix}$$

Chapter 6 Exercise Answers

6.1
The ABD matrix for the $[-45/0/45/90]_s$ laminate;

$$
\begin{bmatrix} A & B \\ B & D \end{bmatrix} =
\begin{bmatrix}
7.41e5 & 2.28e5 & 0 & 0 & 0 & 0 \\
2.28e5 & 7.41e5 & 0 & 0 & 0 & 0 \\
0 & 0 & 2.56e5 & 0 & 0 & 0 \\
0 & 0 & 0 & 4.81e2 & 1.61e2 & -1.04e2 \\
0 & 0 & 0 & 1.61e2 & 2.31e2 & -1.04e2 \\
0 & 0 & 0 & -1.04e2 & -1.04e2 & 1.76e2
\end{bmatrix}
$$

The thermal forces and moments vector for a $[-45/0/45/90]_s$ laminate with material properties from problem 3.1 is;

$$
\left\{ \begin{array}{c} N^T \\ M^T \end{array} \right\}_x =
\left\{ \begin{array}{c}
9.53e-1 \\
9.53e-1 \\
0 \\
0 \\
0 \\
0
\end{array} \right\}
$$

6.2.
The abcd matrix for a $[-45/0/45/90]_s$ laminate with material properties from problem 3.1 is;

$$
\begin{bmatrix}
1.49e-6 & -4.59e-7 & 0 & 0 & 0 & 0 \\
-4.59e-7 & 1.49e-6 & 0 & 0 & 0 & 0 \\
0 & 0 & 3.90e-6 & 0 & 0 & 0 \\
0 & 0 & 0 & 2.77e-3 & -1.62e-3 & 6.78e-4 \\
0 & 0 & 0 & -1.62e-3 & 6.86e-3 & 3.09e-3 \\
0 & 0 & 0 & 6.78e-4 & 3.09e-3 & 7.91e-3
\end{bmatrix}
$$

6.3.

The laminate middle surface stains and curvatures for the boundary condition $Nx = 2400lbs/in$

$$\left\{ \begin{matrix} \varepsilon^o \\ \kappa \end{matrix} \right\}_x = \left\{ \begin{matrix} 3.58e-3 \\ -1.10e-3 \\ 0 \\ 0 \\ 0 \\ 0 \end{matrix} \right\}$$

6.4.

a) For the 90° ply for the boundary condition $Nx = 2400lbs/in$, the homogeneous total strain tensor in the global coordinate system

$$\{\varepsilon\}_{x,90} = \left\{ \begin{matrix} 3.58e-3 \\ -1.10e-3 \\ 0 \end{matrix} \right\}$$

b) For the 90° ply for the boundary condition $Nx = 2400lbs/in$, the homogeneous total strain tensor in the material coordinate system

$$\{\varepsilon\}_{1,90} = \left\{ \begin{matrix} -1.10e-3 \\ 3.58e-3 \\ 0 \end{matrix} \right\}$$

c) For the 90° ply for the boundary condition $Nx = 2400lbs/in$, the homogeneous mechanical strain tensor in the material coordinate system

$$\{\varepsilon\}_{m,1,90} = \left\{ \begin{matrix} -1.10e-3 \\ 3.58e-3 \\ 0 \end{matrix} \right\}$$

d) For the 90° ply for the boundary condition Nx = 2400lbs/in, the homogeneous stress tensor in the material coordinate system

$$\{\sigma\}_{1,90} = \left\{ \begin{array}{c} -2.29e4 \\ 4.25e3 \\ 0 \end{array} \right\}$$

6.5.

The laminate middle surface stains and curvatures for the boundary condition $\Delta T = -100°F$

$$\left\{ \begin{array}{c} \varepsilon^o \\ \kappa \end{array} \right\}_x = \left\{ \begin{array}{c} -9.84e-5 \\ -9.84e-5 \\ 0 \\ 0 \\ 0 \\ 0 \end{array} \right\}$$

6.6.

a) For the 90° ply for the boundary condition $\Delta T = -100°F$, the homogeneous total strain tensor in the global coordinate system

$$\{\varepsilon\}_{x,90} = \left\{ \begin{array}{c} -9.84e-5 \\ -9.84e-5 \\ 0 \end{array} \right\}$$

b) For the 90° ply for the boundary condition $\Delta T = -100°F$, the homogeneous total strain tensor in the material coordinate system

$$\{\varepsilon\}_{1,90} = \left\{ \begin{array}{c} -9.84e-5 \\ -9.84e-5 \\ 0 \end{array} \right\}$$

c) For the 90° ply for the boundary condition $\Delta T = -100°F$, the homogeneous mechanical strain tensor in the material coordinate system

$$\{\varepsilon\}_{m,1,90} = \begin{Bmatrix} -1.28e-4 \\ 1.70e-3 \\ 0 \end{Bmatrix}$$

d) For the 90° ply for the boundary condition $\Delta T = -100°F$, the homogeneous stress tensor in the material coordinate system

$$\{\sigma\}_{1,90} = \begin{Bmatrix} -2.17e3 \\ 2.17e3 \\ 0 \end{Bmatrix}$$

We should expect the 90° ply in the 1-axis fiber direction to be under compression and the 90° ply in the 2-axis transverse matrix direction to be under tension, as is the case above. These values are expected as the 90° ply in the 1-axis fiber direction wants to "shrink" less than all other plies in that direction; therefore, all other plies are trying to "compress" the 90° ply 1-axis fiber direction towards them. The opposite occurs for the 90° ply 2-axis transverse matrix direction. In this direction, the 90° ply 2-axis transverse matrix direction wants to "shrink" more than all other plies in that direction; therefore, all other plies are trying to "pull" the 90° ply 2-axis transverse matrix direction towards them.

6.7.
a) For the 90° ply for the combined boundary condition $Nx = 2400lbs/in$ and $\Delta T = -100°F$, the homogeneous total strain tensor in the global coordinate system

$$\{\varepsilon\}_{x,90} = \begin{Bmatrix} 3.48e-3 \\ -1.20e-3 \\ 0 \end{Bmatrix}$$

b) For the 90° ply for the combined boundary condition Nx = 2400lbs/in and $\Delta T = -100°F$, the homogeneous total strain tensor in the material coordinate system

$$\{\varepsilon\}_{1,90} = \begin{Bmatrix} -1.20e-3 \\ 3.48e-3 \\ 0 \end{Bmatrix}$$

c) For the 90° ply for the combined boundary condition Nx = 2400lbs/in and $\Delta T = -100°F$, the homogeneous mechanical strain tensor in the material coordinate system

$$\{\varepsilon\}_{m,1,90} = \begin{Bmatrix} -1.23e-3 \\ 5.28e-3 \\ 0 \end{Bmatrix}$$

d) For the 90° ply for the combined boundary condition Nx = 2400lbs/in and $\Delta T = -100°F$, the homogeneous stress tensor in the material coordinate system

$$\{\sigma\}_{1,90} = \begin{Bmatrix} -2.51e4 \\ 6.42e3 \\ 0 \end{Bmatrix}$$

Chapter 7 Exercise Answers

7.1.
First the isotropic plane stress stiffness matrix needs to be calculated for the aluminum material;

$$[\overline{Q}] = \begin{bmatrix} 1.12e7 & 3.70e6 & 0 \\ 3.70e6 & 1.12e7 & 0 \\ 0 & 0 & 3.76e6 \end{bmatrix}$$

Then the [A] matrix for the homogeneous isotropic aluminum plate is;

$$[A] = \begin{bmatrix} 8.98e5 & 2.96e5 & 0 \\ 2.96e5 & 8.98e5 & 0 \\ 0 & 0 & 3.01e5 \end{bmatrix}$$

And the [D] matrix for the homogeneous isotropic aluminum plate is;

$$[D] = \begin{bmatrix} 4.79e2 & 1.58e2 & 0 \\ 1.58e2 & 4.79e2 & 0 \\ 0 & 0 & 1.60e2 \end{bmatrix}$$

Finally, the [B] matrix equals zero for any homogeneous plate; [B] = 0

7.2.
The equivalent in-plane homogenized engineering constants for the [−45/0/45/90]$_s$ laminate;

$E_x = 8.38e6$
$E_y = 8.38e6$
$v_{xy} = 0.308$
$G_{xy} = 3.21e6$

Notice that the equivalent in-plane homogeneous engineering constants for a [−45/0/45/90]$_s$ laminate produces isotropic engineering constants; $E_x = E_y$ and $G_{xy} = E_x / (2 (1+v))$. Furthermore, this laminate is balanced and symmetric. Due to these facts, balanced and symmetric laminates are often referred to as quasi-isotropic laminates.

7.3.

The equivalent bending homogenized engineering constants for the $[-45/0/45/90]_s$ laminate;

$E_x = 8.46e6$
$E_y = 3.42e6$
$v_{xy} = 0.586$
$G_{xy.} = 2.96e6$

7.4.

The equivalent material matrices for the $[-45/0/45/90]_s$ laminate;

$$[\overline{Q}_1] = \begin{bmatrix} 9.26e6 & 2.85e6 & 0 \\ 2.85e6 & 9.26e6 & 0 \\ 0 & 0 & 3.21e6 \end{bmatrix}$$

$$[\overline{Q}_2] = \begin{bmatrix} 1.13e7 & 3.77e6 & -2.44e6 \\ 3.77e6 & 5.41e6 & -2.44e6 \\ -2.44e6 & -2.44e6 & 4.13e6 \end{bmatrix}$$

$$[\overline{Q}_4] = \begin{bmatrix} 0 & 0 & 0 \\ 0 & 0 & 0 \\ 0 & 0 & 0 \end{bmatrix}$$

7.5.

The equivalent transverse shear material matrix for a $[-45/0/45/90]_s$ laminate with material properties from problem 3.1 is;

$$[\overline{Q}_3] = \begin{bmatrix} 5.84e5 & 0 \\ 0 & 6.34e5 \end{bmatrix}$$

7.6.

Given the ply properties calculate $\overline{Q}_{11}$ and $\overline{Q}_{44}$ for the 0, 90, 45, and -45 plies;

$\overline{Q}_{11,0} = 22.12e6$

$\overline{Q}_{11,90} = 1.31e6$

$\overline{Q}_{11,45} = 6.80e6$

$\overline{Q}_{11,-45} = 6.80e6$

$\overline{Q}_{44,0} = 0.75e6$

$$\overline{Q}_{44,90} = 0.75e6$$
$$\overline{Q}_{44,45} = 5.66e6$$
$$\overline{Q}_{44,-45} = 5.66e6$$

Given the aluminum properties and plate thickness along with the $\overline{Q}_{11}$ and $\overline{Q}_{44}$ components above;

$$\begin{bmatrix} 2.34e7 & 1.36e7 \\ 1.50e6 & 1.13e7 \end{bmatrix} \begin{Bmatrix} T_0 \\ T_{45} \end{Bmatrix} = \begin{Bmatrix} 8.98e5 \\ 3.01e5 \end{Bmatrix}$$

Inverting to solve for T_0 and T_{45};

$$\begin{Bmatrix} T_0 \\ T_{45} \end{Bmatrix} = \begin{bmatrix} 4.62e-8 & -5.56e-8 \\ -6.13e-9 & 9.57e-8 \end{bmatrix} \begin{Bmatrix} 8.98e5 \\ 3.01e5 \end{Bmatrix}$$

$T_0 = T_{90} = 0.0248$"
$T_{45} = T_{-45} = 0.0233$"
Total Laminate Thickness = 0.0962". If each ply was 0.01" thick, then you would need at least 4 plies of each angle to approximate the behavior of the aluminum plate. A possible stacking sequence would be [-45/0/45/90/0/90/45/0/-45].

7.7.
Given the ply properties calculate $\overline{Q}_{11}$ for the 0, 90, 45, and -45 plies;
$$\overline{Q}_{11,0} = 22.12e6$$
$$\overline{Q}_{11,90} = 1.31e6$$
$$\overline{Q}_{11,45} = 6.80e6$$
$$\overline{Q}_{11,-45} = 6.80e6$$

Given the aluminum properties and plate thickness along with the $\overline{Q}_{11}$ components above;

$$T_0 = \sqrt[3]{\frac{(10e6)(0.08)}{(16)(1-0.33^2)(3.70e7)}}$$

$T_0 = T_{90} = T_{45} = T_{-45} = 0.0213$
Total Laminate Thickness = 0.0853". If each ply was 0.01" thick, a possible stacking sequence would be [-45/0/45/90]$_s$

Chapter 8 Exercise Answers

8.1.
The ABD matrices for the various laminates are;
a) $[0/45/90/-45]_s$

$$
\begin{bmatrix} A & B \\ B & D \end{bmatrix} =
\begin{bmatrix}
7.41e5 & 2.28e5 & 0 & 0 & 0 & 0 \\
2.28e5 & 7.41e5 & 0 & 0 & 0 & 0 \\
0 & 0 & 2.56e5 & 0 & 0 & 0 \\
0 & 0 & 0 & 6.42e2 & 8.22e1 & 6.24e1 \\
0 & 0 & 0 & 8.22e1 & 2.26e2 & 6.24e1 \\
0 & 0 & 0 & 6.24e1 & 6.24e1 & 9.75e1
\end{bmatrix}
$$

b) $[45/90/-45/0]_s$

$$
\begin{bmatrix} A & B \\ B & D \end{bmatrix} =
\begin{bmatrix}
7.41e5 & 2.28e5 & 0 & 0 & 0 & 0 \\
2.28e5 & 7.41e5 & 0 & 0 & 0 & 0 \\
0 & 0 & 2.56e5 & 0 & 0 & 0 \\
0 & 0 & 0 & 2.31e2 & 1.61e2 & 1.04e2 \\
0 & 0 & 0 & 1.61e2 & 4.81e2 & 1.04e2 \\
0 & 0 & 0 & 1.04e2 & 1.04e2 & 1.76e2
\end{bmatrix}
$$

c) $[90/-45/0/45]_s$

$$
\begin{bmatrix} A & B \\ B & D \end{bmatrix} =
\begin{bmatrix}
7.41e5 & 2.28e5 & 0 & 0 & 0 & 0 \\
2.28e5 & 7.41e5 & 0 & 0 & 0 & 0 \\
0 & 0 & 2.56e5 & 0 & 0 & 0 \\
0 & 0 & 0 & 2.26e2 & 8.22e1 & -6.24e1 \\
0 & 0 & 0 & 8.22e1 & 6.42e2 & -6.24e1 \\
0 & 0 & 0 & -6.24e1 & -6.24e1 & 9.75e1
\end{bmatrix}
$$

d) $[-45/0/45/90]_s$

$$\begin{bmatrix} A & B \\ B & D \end{bmatrix} = \begin{bmatrix} 7.41e5 & 2.28e5 & 0 & 0 & 0 & 0 \\ 2.28e5 & 7.41e5 & 0 & 0 & 0 & 0 \\ 0 & 0 & 2.56e5 & 0 & 0 & 0 \\ 0 & 0 & 0 & 4.81e2 & 1.61e2 & -1.04e2 \\ 0 & 0 & 0 & 1.61e2 & 2.31e2 & -1.04e2 \\ 0 & 0 & 0 & -1.04e2 & -1.04e2 & 1.76e2 \end{bmatrix}$$

The [A] matrices for the various laminates are all the same. This should be the case as the [A] matrix is stacking sequence independent. The [D] matrices for the various laminates are all different. This should also be the case as the [D] matrix is stacking sequence dependent.

8.2.
We notice that $[D]_{average} = [D]_{smear}$. This shows that $[D]_{smear}$ is the average [D] matrix for all possible stacking sequences of a given laminate.

$$[D]_{average} = \begin{bmatrix} 3.95e2 & 1.21e2 & 0 \\ 1.21e2 & 3.95e2 & 0 \\ 0 & 0 & 1.37e2 \end{bmatrix}$$

$$[D]_{smear} = \begin{bmatrix} 3.95e2 & 1.21e2 & 0 \\ 1.21e2 & 3.95e2 & 0 \\ 0 & 0 & 1.37e2 \end{bmatrix}$$

8.3.
$[D]_{smcore}$ for the laminates with a core layer of thickness $t_c = 0.25$ inches is;

$$[D]_{smcore} = \begin{bmatrix} 1.57e4 & 4.82e3 & 0 \\ 4.82e3 & 1.57e4 & 0 \\ 0 & 0 & 5.43e3 \end{bmatrix}$$

8.4.

a) True; the [A] matrix is stacking sequence independent.

b) False; in order for D_{14} and D_{24} to equal zero the laminate must be balanced and anti-symmetric.

c) False; the equivalent in-plane homogenized axial stiffness E_x is the same for both laminates. Any n multiple of a given laminate thickness will produce the same equivalent in-plane homogenized engineering constants.

d) False; D_{11} and D_{22} will always be different for a symmetric laminate, unless it is the trivial symmetric laminate of $[\theta]_n$

8.5.

a) $E_x = E_y$; Since the laminates are the same in the x- and y-directions, therefore $A_{11} = A_{22}$.

b) $D_{11} > D_{22}$; Since the $0°$ ply is further away from the neutral axis in the x-direction and the $0°$ ply is one ply closer to the neutral axis in the y-direction.

c) $A_{14} = 0$; Since the laminate is balanced.

8.6.

a) $E_x > E_y$; Since there are two $0°$ plies in the x-direction and only 1 $0°$ ply in the y-direction, therefore $A_{11} > A_{22}$.

b) $D_{11} > D_{22}$; Since the $0°$ plies are further away from the neutral axis in the x-direction and the one $0°$ ply is two plies closer to the neutral axis in the y-direction.

c) $A_{14} = 0$; Since the laminate is balanced.

8.7.

The stacking sequence to get the maximum D_{44} component is shown below.

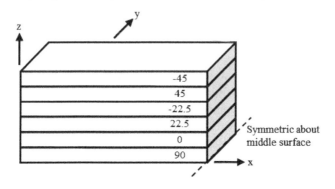

8.8.

We should add a 22.5°, 45°, and −30° plies to the laminate in order to balance the laminate. Balanced laminates do not exhibit extension-shear coupling due to $A_{14} = A_{24} = 0$.

8.9.

Existence of the D_{14} term would cause a laminate to twist under a pure bending load M_x. This behavior should be more evident in thin laminates. In general, for laminates greater than 16 plies, D_{14} and D_{24} components become small compared to the other [D] matrix components and the bending – twist coupling which results from the D_{14} and D_{24} components is negligible.

8.10.

$D_{11,A}$ is on the order of $37Q_{11}$. $D_{11,B}$ is on the order of $7Q_{11}$. Therefore the ratio of $D_{11,A}/D_{11,B} = 37/7$ or 5.28 times larger.

8.11.

Put an X in the boxes that apply for each laminate. Assume same material and thickness for each ply.

Laminate	Symmetric	$A_{14} = A_{24} = 0$	$D_{14} = D_{24} = 0$
[0/90/0/90/0]	X	X	X
[45/−45/−30/30]		X	
[0/45/90/45/0]	X		
[45/−45/90/0/90/−45/45]	X	X	
[0/90/−45/45/90/0]		X	X
[90/0/90/0]		X	X
[0/45/−45/0]		X	X
[22.5/−22.5/90/−22.5/22.5]	X	X	
[−45/22.5/0/22.5/−45]	X		
[−30/30/60/30/−30]	X		

Chapter 9 Exercise Answers

9.1.
The Tsai-Wu first ply failure index for $N_x = 2400$lbs/in is;
FI = 0.62

9.2.
The Tsai-Wu first ply failure index for $\Delta T = -100°F$ is;
FI = 0.28

9.3.
The Tsai-Wu first ply failure index for the combined boundary condition $N_x = 2400$lbs/in and $\Delta T = -100°F$ is;
FI = 1.01 (90° ply failure)

Chapter 10 Exercise Answers

See appendix E.

Chapter 11 Exercise Answers

See appendix F.

References

1. Altair Engineering, Inc. *OptiStruct Reference Guide.* Troy, MI, 2011.
2. Arora, Jasbir S. *Introduction to Optimum Design.* San Diego, CA: Elsevier Academic Press, 2004
3. Chou, Pei Chi, and Nicholas J. Pagano. *Elasticity: Tensor, Dyadic, and Engineering Approaches.* New York: Dover Publications, 1967.
4. Jones, Robert M. *Mechanics of Composite Materials.* 2nd ed. Philadelphia: Taylor & Francis, 1999.
5. Schramm, Uwe, and Harold Thomas, Ming Zhou. *Issues of commercial optimization software development.* Structures and Multi-disciplinary Optimization 23, 97-110: Springer-Verlag, 2002.
6. Timoshenko, Stephen. *History of Strength of Materials.* New York: Dover Publications, 1983.
7. Tsai, Stephen W., and Thomas H. Hahn. *Introduction to Composite Materials.* Westport, CT: Technomic Publishing, 1980.
8. Ugural, A. C., and Saul K. Fenster. *Advanced Strength and Applied Elasticity.* Upper Saddle River, NJ: Prentice-Hall, 1995.

Made in the USA
Las Vegas, NV
07 July 2021